The Seven Modules of Systematic Maintenance

The Seven Modules of Systematic Maintenance

Systematic Maintenance Technologies

L. Robert Pyle

iUniverse, Inc.
New York Lincoln Shanghai

The Seven Modules of Systematic Maintenance

All Rights Reserved © 2003 by Lauri R. Pyle

No part of this book may be reproduced or transmitted in any form or by any means, graphic, electronic, or mechanical, including photocopying, recording, taping, or by any information storage retrieval system, without the written permission of the publisher.

iUniverse, Inc.

For information address:
iUniverse, Inc.
2021 Pine Lake Road, Suite 100
Lincoln, NE 68512
www.iuniverse.com

ISBN: 0-595-30421-4

Printed in the United States of America

Contents

An Introduction vii
Maintenance Operating Module 1
Maintenance Management Module 11
Maintenance Reliability Systems 22
Maintenance Planning Module 38
Maintenance Efficiency Module 46
Maintenance Training Module 60
Maintenance Safety Module 66
Appendix A 79
Appendix B 83

An Introduction

Have you tried the advanced Maintenance Theories only to find them too complicated or impossible to sustain? Have you gone to one of those seminars and come away feeling confused and discouraged. The problem is probably not in those theories but in your basic maintenance systems.

All advanced systems depend on having a foundation of maintenance systems in place, even if they won't admit that. Without an operating system, a management system, a planning system, some metrics to measure progress, etc, you are trying to build a house on a sand foundation. Scripture tells us what happens to a place like that the first time a storm comes along.

Just as every physical structure needs a strong foundation, every organizational structure must be built on foundational principles. Systemaitec has produced the Seven Modules of Systematic Maintenance as foundation stones on which to build a truly excellent maintenance organization.

Systemaitec's seven modules are not magic. They don't produce miracles—well, maybe they do, when applied correctly to disorganized maintenance efforts. Without these seven foundational blocks in place, though, your maintenance function will never be all that it should be. Putting these modules in place or tuning up the ones you have but have ignored for so long will not suddenly enable you to cut your maintenance crew and double production. They will allow you to continuously improve reliability, costs, and production equipment on-stream. They are not silver bullets. They are just common sense. They aren't foo-foo dust that will suddenly make you the apple of management's eye. They require hard work, discipline and consistency. With those three resources applied, you will see the performance improvements that you have been after. Then you can take all that information you got at the last seminar and apply it productively to an already top-notch department.

The Seven Modules

MOS—The Maintenance Operating System

This module sets up the workflow and the forms needed for getting work into the department from customers, processing the work to make its execution by craftspeople efficient, prioritizing the work, scheduling the work, closing out the work, and reporting the completion to the customer.

MMS—The Maintenance Management System

This module sets up the systems needed to plan, organize, and control the maintenance department. It sets up equipment codes, sets up systems to capture and report costs, determines how much PM and PdM is economical, and gives the manager the tools needed to manage and measure the department's performance.

MRS—The Maintenance Reliability System

This module sets up the PM and PdM programs that are needed to lower costs and improve equipment reliability. This is not a "one size fits all" approach. These programs are customized for your plant.

MPS—The Maintenance Planning System

This module determines the level of planning that is right for your situation. Once that is determined, it sets up the systems you need to get that planning done. Again, this is a customized module. You only get the systems that will work for you and that you need.

MES—The Maintenance Efficiency Module

This module established metrics for maintenance performance and methods for tracking, reporting and improving them. Maintenance costs, labor usage, labor effectiveness, and system effectiveness are all measured from information that is mostly already available in your maintenance department.

MTS—The Maintenance Training System

How do you hire craftspeople? How do you know you are getting what you expect? How do you insure your crafts are maintaining their skills so that they are going to be able to handle future work and future technology? This module answers those questions for you. It will set up systems that tie you into the mainstream of training.

MSS—The Maintenance Safety System

Craftspeople are hard to come by. Losing one of them, even for a few days, to an injury puts a large strain on your resources. This module sets up a basic safety program that will start you on your way to attaining a zero Lost Time Injury rate. This program, based on very successful models practiced throughout industry, is customized for your situation.

Systemaitec has developed this unique modular concept to allow you to get what you need to advance to the next level in maintenance. If you already have a work order system but lack the MMS module, we can modify our module to interact with your existing system. If you have a great department but need a good safety system, the MSS will fit into any department. If you simply need metrics, we can establish and track those for you. These modules, individually or corporately, can be set up in conjunction with your existing system.

The MOS and MMS are so basic that if you did not have these systems in some form already in your department, we would insist that you install ours before trying to put anything else in place. Without a basic management scheme and a basic operating scheme in place, nothing else will work.

In the rest of this book, you will find a detailed description of these modules. Included are examples of the forms we can supply to you, modified for you company name and logo. Additionally, you will find descriptions of what Systemaitec will do to install each module with options that you may choose to add for an additional charge.

Maintenance Operating Module

Maintenance departments are service organizations. They do not make product for the customer. They have nothing to sell to the outside world. What they have is labor and expertise to sell to the rest of the internal organization. Because of that, Maintenance normally is not the master of its own fate. Maintenance depends on the rest of the organization, normally the production department, to provide the work that Maintenance will perform.

That does not mean that the Maintenance department has no work of its own. Nor does it mean that the Maintenance department is at the whim of the production people. It does mean that Maintenance must meet the needs of its customers and has to have a means of knowing what it is the customer wants, how to prioritize that into the overall scheme of work, and how to schedule with its customers to make the equipment and facilities available to get that work done.

CMM Systems (Computerized Maintenance Management Systems)

There are excellent CMM systems on the market. Some basic systems sell for around one thousand dollars. Some very popular systems that can link multiple sites, purchasing, stores, PM, and Work Request can sell for as much as $15,000 per license. Which is right for you? The one criteria you cannot use is price. I have installed two thousand dollars systems that had functionality for that client that the $15,000 system could not match. In other situations, only the high price systems would do everything the client wanted. There are a few things, though, to look at when buying a system:

1. Make sure it has all the modules included in the base price. Many vendors will sell you the basic system. It is after you have it and want to add on the stores module that you find you are stuck for another $8,000 charge and some programming time.

2. Make sure it interfaces with your existing accounting and purchasing systems. There are many high priced systems that are stand-alone. The developers used their own database structure and you will not be able to get into or out of that database without a patch (an expensive one) from them. This is where a cheaper system that uses a standard database driver might be much more economical. Normally, your own IT can write tie-programs to make the two or three systems you are operating talk to one another.

   One client of mine purchased a $1500 system that was complete in itself. It could do stores, purchasing, and all the maintenance functions. Unfortunately (and is almost always the case,) the company already had an inventory system and a purchasing system that was already tied into their accounting system. So, my job was to find a way to make it work. Knowing these things prior to the purchase allowed me to pick a relatively inexpensive but robust system that used a very common database structure. Their own IT wrote ties to purchasing and stores and the system was up and running in less than a month.

3. Make sure that there is a separate module that allows your internal maintenance customers to write work requests without having to have access to the whole program—which can give them access to parts of the database you want to keep restricted and can be confusing to the operator on the Weber that just wants to ask for somebody to stop the growl in the bearing. Many expensive systems do not allow for this function and you end up using expensive seats just to allow people to fill out a work request. At one client's plant, the planner was locked off the system for a couple of hours because too many people had logged in at once. Since he was the only one in Maintenance trying to log in, it was obvious that some operators had logged in to write work requests and

had failed to log back out. Look for a system that gives you unlimited or at least a very large number of seats for creating work requests.

The Maintenance Work Order

See the Form in Appendix A

The most important system for communicating between Maintenance and its customers is the Work Order System. Most companies have converted to a CMMS. The form of the work order in a CMMS is normally set by the programmer. It is our experience that these work orders normally conform to the standards described below. If you use a CMMS, read this section as a primer on why the work order looks like it does.

If a computerized system is being used, data is all kept on a server so a paper form is not necessary. I have found, though, that production people want the feel of a piece of paper in a file or to have access to the backlog in the CMMS. Good operations managers and supervisors who have accepted ownership of the equipment they operate want that sense of security about what has been ordered and what has not been ordered. If you implement an electronic work order system, have a way for the originator and the authorizer to print a hard copy of the request with the identifying number on it and give them access to the database—at least, read-only access (You can add this to the things a good CMM system should do.)

If you are using a paper system, which I recommend for the organization that has never had a good work order system in place, use a multi-copy form on NCR paper. The number of copies depends on how you are organized. At a minimum, it should be two copies—one for the originator/authorizer and one for the maintenance department. I recommend three copies: one for the originator, one for the maintenance foreman/planner, and one to give to the lead craftsperson. This insures that a good record can be kept even if one copy is lost during the performance of the job.

Tracking

The work request should be pre-numbered—whether they are electronic or paper. This allows the form to be tracked even before it shows up at the Maintenance department. These should be at least six-digit, alphanumeric identifiers. That insures that you can write about one and a half trillion work orders before having to re-use numbers. I don't honestly think you will do that, but you don't want to take the chance of having to ever reuse an identifier. Computers just don't understand.

Why alphanumeric? If you use straight numbers, the most unique numbers you can obtain from six digits is one million. That may be plenty, but why take the chance?

Every work order should have the equipment code and the name of the equipment written on it. In the Maintenance Management Module, we will discuss the equipment code, why it is absolutely necessary, how to assign it, and what format to use. For now, suffice it to say that tracking the work order, finding the equipment, identifying spare parts, creating an equipment history, and charging labor and supplies will all depend on that number. If you are implementing a CMMS nothing else can happen until that number is assigned.

Every work order should have the accounting number where the requesting department wants the labor and parts charged. This may be filled in automatically if you have a CMMS since that number is linked to the equipment code. Even on paper, the normal charge would be to the equipment code with accounting then assigning the costs to the linked department. There is some work, though, where the equipment should not be charged—such as major maintenance work or equipment modification or adjustment or support for a production experiment. In those cases, it is necessary for the proper charge number to be entered by the requesting department. Your CMMS system should have the ability to charge to both an equipment number or to a charge code. Oddly, some of the very good systems on the market do not give you this ability. If you have the option to change a system that does not have that ability, then exercise it. If you are writing a specification for a new CMMS, make sure that you include that.

All requests should be dated when filled out. This allows the Maintenance department to track how well it is completing requested work through an "aging days" report. We will discuss this report in the Maintenance Efficiency Module.

Priority

When do you schedule a maintenance work request? There are many answers for this one. The first step in determining that date is to assign priority to the work. Priority lets the maintenance department know how important the production department thinks the work is. Many maintenance consultants utilize the 4 Priority system. This uses only four codes as follows:

Emergency—drop all other work and begin this immediately.

Urgent—must be started within 24 hours.

Routine—should be kept in backlog and scheduled as parts, materials, and the equipment become available.

Shutdown—schedule during planned equipment outages only.

Some modify this with a fifth priority called safety but it is just a subset of emergency. Still others have modified it with a Safety 1 and Safety 2 system that are subsets of Emergency and Urgent.

Every place I have seen this priority system used, I have found informal priority classifications that supercede it. The reason for this corruption of the system is that these four classifications are too gross to express all the subtleties of the interaction necessary between Maintenance and Production to effectively get all the Maintenance duties accomplished and still satisfy all of Production's needs.

I recommend a different, more complex but easy to use system. It is built on the original Rime system that was standard in the sixties and seventies. The codes and definitions are listed at the end of this module.

It is absolutely necessary for production to have a major hand in setting the priority of the work. It is their equipment you are maintaining. They are the customers. On the other hand, maintenance has the expertise in maintaining and repairing equipment. They must have an input on how they apply limited resources to accomplish all the customer wants and still get all of their other work accomplished. This priority system provides that input from both groups.

Production should assign their code in the proper box prior to submitting the work request. The maintenance planner/foreman/superintendent assigns the maintenance code. The two codes are multiplied together. The result is a priority number from 3 to 90—90 being the most important. (see the definitions in Appendix A) Translating these numbers into the 4 Priority system.

90	Emergency	Begin immediately
89–70	Urgent	Begin as soon as possible.
70–3	Routine	Backlog work.

What do you want done?

There are at least four different types of work request. One requests maintenance work on equipment that has failed. The other requests maintenance work of equipment that is still running but which is not running as expected or not producing product that meets specifications. The third requests maintenance work that has nothing to do with equipment repair, such as facility upkeep, modifications, or installation of new equipment. The fourth requests Preventive or Predictive Maintenance to be done on plant equipment or facilities. This type is normally an internal maintenance work order.

When the request is of the first two types, it may be advantageous for the production department to fill out the section on symptoms. If the work needed is not as obvious as a frozen bearing or a broken strut, and especially if the equipment is still running but not doing what is expected, a list of symptoms by the operator will improve maintenance repair time by providing needed information up front. Manuals, tools, and test equipment needed to troubleshoot can be brought to the job site initially instead of having to return to the files or the tool room after the job is started.

If the work is of the third type, then nothing is needed in the symptoms section.

The description of requested action should be complete enough that the planner and/or craftsperson do not have to hunt up the requester to find out what is needed. Recycling information is a waste of time on everyone's part. Production should ensure there is a clear description of what needs to be done and what the desired outcome is. They should attach sketches, mechanical drawings, descriptions, pages from a manual, pictures, or whatever else that will make the scope of the job clear to the maintenance department.

Who wants it done?

Sometimes, no matter how good the description is, the craftsperson or planner is going to need more information. The requester should sign and date the work request. This gives the planner or craftsperson a contact point. Also, it lets the Maintenance department know whom to notify when the work is complete. A legible signature is a real benefit to efficiency and planning. It does no good to save an hour of craft time by planning the job if you lose an hour of craft or planning time trying to figure out who requested the work.

Obviously, CMMS eliminates this problem by recording who initiated the work order automatically. If you are using a CMMS, then ensure that the system is set up to record who initiated the work order, not whose computer the work request was submitted from.

I had one client who had two problems with this. They wanted anyone on the line to be able to use the line computer to submit a request. They also wanted anyone on the line to be able to view all the work orders submitted for that line so that duplication was eliminated. The computer assigned whoever was logged into the computer as the initiator. In order to accomplish all they wanted without the hassle of signing on and off each user each time and to make it easy to view all the work submitted for any line, we made the line the user. We signed the line onto the CMMS and told each user to put their names in the description of the work to be performed. This way, maintenance knew who to ask and everyone on the line had access to every work order that was initiated.

When Maintenance gets a work request, they are going to spend someone else's money. Some managers don't see it that way, but it is true. Many Maintenance departments have charge-out rates in the low $100's. Some smaller departments may direct charge the labor and parts. This can still add up to a great deal of money. With a request in hand, Maintenance is authorized to spend a lot of somebody else's funds. Shouldn't someone who has responsibility for those funds—who will have to answer at month's end for the expenditures—know that the money is being spent? The answer is obviously: yes. That is why we recommend that the budget manager authorize all maintenance work. This can be delegated to a foreman or day foremen but should not be simply left up to the discretion of the machine operator. Although they may have the best intentions, they don't have the knowledge of the money that is being spent elsewhere—and therefore, can ask for work that is not needed that month or unwanted by the department management.

What is needed to get the job done?

Systemaitec believes that all work should be planned before assigning the work to a craftsperson. This planning involves parts, tools, equipment, and labor. We will cover this in the Maintenance Planning Module. For now, the work order form is set up to record all that planning activity. This is not just a convenient place for the planner to doodle. This form is the main vehicle for getting job information from the production department to the craftsperson assigned to the work. Recording or attaching all planning information insures that the one performing the work is made aware of all the aspects of the job: tools needed, equipment needed, parts needed, other crafts needed, and how long the job is expected to take. All of that information goes to the job site with the craftsperson.

Was the job completed the way it was described?

The Maintenance department often encounters problems, failures, or other conditions that were not anticipated in the original request. That information is utterly important for improving the efficiency of the department. It is also important for the engineers who are going to be ordering new equipment in the future.

The completed work section of the request is the opportunity for the craftspeople to record what they found, what they did, and what other information they feel should have been transmitted to them before the beginning of the job. This is especially important when the corrective action did not coincide with the Production department's initial idea of what was wrong.

This should be filled out by the craftsperson. Many departments relieve the field people of this duty, leaving it to the foreman to complete. This approach is guaranteed to lose important information. Every maintenance foreman should insist that this section be completed prior to receiving the work order back.

When the foreman assigns the work, he/she should date the request in the Date Work Started block. The craftsperson should sign and date the work request when the work is completed. The signature should be legible. Again, losing engineering or supervisory time hunting down who signed a work order is not a productive use of limited resources.

Records

When complete, the work order is a historical document. It contains important information about the equipment, about the job, and about the planning done that will be very important in any department's efforts to improve. We will be covering some of this information in the Maintenance Efficiency Module. In order for this information to be useful, it has to be retrievable. To be retrievable, it has to be filed properly.

If you are using CMMS, then the data have to be taken from the printed form and entered into the database. That is extremely tedious and it is labor intensive. The results, though, are data that are easy to manipulate, retrieve and study.

If you are not using CMMS, then I recommend that you establish a filing systems based on the equipment code—one hanging folder per piece of equipment. Every work order with attached documentation should be filed by code chronologically. This results in a system into which data are easy to enter but are harder to retrieve. Of course, for a small manufacturer with a limited amount of equipment, this may be the simpler, cheaper, and more efficient system to use. Whether computerized or manual, this system can be extremely useful in identifying improvements and changes needed to lower maintenance costs. The maintenance manager or engineer should review these files annually to identify needed changes to equipment, procedures, tooling, or equipment.

Work Flow

See the flow sheet in Appendix A

All work other than preventive and predictive maintenance duties comes from departments outside of Maintenance. In the Systemaitec Maintenance Operating Module, this is the workflow of a work request:

1. The initiator fills out all the shaded areas on the request except the Production code and the authorizer's signature. The initiator then sends all copies to the authorizer. In a CMMS, this is done by simply saving the work request.

2. The authorizer checks the order for completeness and appropriateness. If the work is approved, the authorizer fills in the production code portion of the priority and signs the form. The authorizer files the bottom copy of the form and forwards the other two copies to the maintenance department. This is accomplished in CMMS by saving the request.

3. The maintenance foreman/planner receives the order and records its receipt in a log. This may be part of a CMMS system or a manual log. At this time, he/she assigns the maintenance code and calculates the priority. Emergency orders are sent to the foreman for immediate assignment of resources. If the order is not an emergency (below priority 90), the planner/foreman does not forward the order until all planning has been complete. As necessary, the planner/foreman investigates the scope of the job to assign resources, obtains parts and special tools, and puts the order in the active backlog.

4. Weekly, Maintenance and Production department representatives meet to plan the next week's work. All work orders in the active backlog are presented to Production as candidates for work. Between the two departments, work is assigned by priority and aging days to utilize all the resources available. This schedule is published by the Maintenance department to all the other support and operating departments.

5. Daily, either before the day shift or at the end of the day shift, the Maintenance foreman/planner and a representative of the Production department meet to discuss changes that may be needed in the next day's schedule.

6. At the beginning of each shift, the maintenance foreman assigns work to each craftsperson according to the schedule and plan provided. The craftsperson receives the MWO and all attachments, including planning documents.

7. The craftsperson completes the assigned work, fills out the completed work section, and returns the package to the foreman. The foreman collects all documentation and returns it to the planner.

8. The planner examines all returned documents for completeness and enters the completed data in the database. The original work order and all attached documents are filed in the history file. A copy of the work order, but not the attached documentation, is sent to the authorizer to notify them that the work is complete. In most CMMS, this is automatic

Using this operating system, you will be collecting all the information you need to create a great management system. If you skip any of these parts, your management system is going to suffer down the road. Set up your operating system first—and apply the discipline to it to make it work as I have explained in this module. You'll be glad you did.

Maintenance Management Module

This module installs the systems needed to manage (plan, organize and control) the maintenance function. Maintenance is not just a few guys turning wrenches. Maintenance is normally a significant portion of the operating budget—perhaps 10% to 15% of what is spent each year. With that much money passing through this organization, it is absolutely necessary to apply the same level of management that one would apply to production.

In this module, you will install systems to track your costs, track your crafts' performance and identify areas of opportunity. We will be covering many of the measurements you could and should make in the Maintenance Efficiency Module. For now, the systems that need to be in place to make those measurements are our concern.

The Equipment Code

The heart of a Maintenance Management System (MMS) or a Computerized Maintenance Management System (CMMS) is the equipment code. Until you have assigned a unique, rational code to each piece of equipment, building, structure, and vehicle in your company, you will have an ineffective management system. Below you will find detailed instructions for assigning these equipment codes. The system proposed by Systemaitec and most other consultants has the advantage of identifying not only the equipment but its type, location and account number. With only a glance at the work order, your planner, maintenance foreman, and craftsperson will know that the equipment to be worked on is, for example, a piston compressor located in the compressed air section of the utility area, or perhaps, for example, a positive displacement pump located on the #2 Boiler.

Capturing Costs

Once the code scheme has been fully implemented, you will have in place all you need to insure that all parts and labor are charged to the exact piece of equipment consuming them, the parts ordered from stores are delivered in boxes marked specifically for the job you have planned and that purchase orders are charged to the accounts that are actually using the purchased material. This, of course, will mean working closely with your accounting department and purchasing department to set up your codes in their systems. If you are worried about getting their cooperation, I have never found these groups to be reluctant to add more detail and better control to expenditures and parts.

Capturing Downtime

Working with the operating department, downtime can now be recorded against the exact piece of equipment that is causing it. No longer will the maintenance manager have to guess as to what caused all the downtime in the previous month, quarter or year. The equipment can be identified on a minute to minute basis and the history file is there to tell him/her what the problem was.

Using the Data

With all work orders and downtime charged to the equipment, with all labor, parts and materials charged (either through the work order number or through a labor and parts distribution accounting system) to the equipment, the maintenance manager is now armed with the most formidable management tools available.

Either manual or computer reports should be generated monthly, quarterly, and annually to identify high cost equipment and equipment that is causing the most downtime for production. Quarterly downtime and spending trends should be charted to identify improving or worsening situations. With these numbers in hand, the maintenance department can apply their limited resources in the most effective manner to eliminate waste, improve on-stream, and modify problem equipment. With these reports, justification for maintenance-generated capital spending is no longer guesswork and good fiction writing but factual and realistic.

Systemaitec recommends that the maintenance manager generate two top ten lists. The first is the equipment that is costing the most to maintain. Resources should then be applied based on the ranking on the list to investigate why the equipment is costing so much and to make recommendations of changes to lower those costs. The second is the equipment that is causing the most downtime. Again, resources should be applied to identify and correct the causes—based on the ranking on the list.

We recommend other reports be generated annually: the cost of PM/PdM for each piece of equipment and for the overall plant; the cost of PM/PdM, Emergency, Urgent and Routine work, the percent of total work hours for PM/PdM, Emergency, Urgent and Routine work. These can be generated because you have the work order numbers, the priority, and the equipment numbers. Ask your accountant what you need to do to get these reports.

With the above reports, you can manage the amount of PM/PdM to minimize costs. There is a level of PM where the cost of the preventive effort outweighs the savings in increased on-stream. Although this is absolutely true for every industry and company, where the economic point is cannot be generalized. A study must be done—with real data (See below on Plotting the Economic Point.)

Systemaitec recommends that all data being reported on completed work orders be filed by equipment code. This includes: how long the job took to complete, the number of craftspeople required to complete it, special tools needed, parts lists, and any special conditions encountered on that particular job. CMM Systems normally record and sort this information automatically. Manual systems must be set up to gather this information. Alternatively, the maintenance department can sort through the filed work orders on a particular piece of equipment on a periodic basis and generate this data. For small organizations with a minimum amount of equipment, this may be the most economical method for gathering and using the data.

With this data, the maintenance manager can generate "normal" times for repetitive tasks or repairs. This data can be used by the planner for future work and can be used by the manager to gage the level of performance of the assigned technicians. Also, using good industrial engineering techniques, resources can be applied to continuously improving these numbers making the department and each individual crew more efficient.

Plotting the Economic Point

Predictive maintenance does not lower maintenance costs. It may even raise maintenance cost through increased capital and labor. It may save parts costs by alerting maintenance to component failures before they become catastrophic, damaging more of the machine, but this is normally at best a break even benefit. The real savings in PdM is in reducing downtime. With good PdM, remaining life can be predicted so that repairs are made when the production equipment is not needed. The job can be well planned and parts obtained on a scheduled rather than emergency basis. The job will go much quicker than if it were done as an emergency.

Preventive Maintenance does lower overall, long term maintenance costs. In the short term, though, it costs money to operate. The costs are now. The benefits may be years in the future.

Most accounting departments can tell you what an hour of production time is worth. This number is used to calculate the value of the downtime lost to equipment failure and maintenance.

Plot a graph with the Y axis plotting dollars and the X axis plotting the number of hours or PM and PdM being done. Plot two curves on the graph. The first is the PM and PdM cost at each level of effort. This should include capital depreciation for equipment purchased such as vibration analyzers or lubrication analyzers, labor, and parts used for PM purposes. 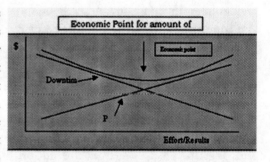 The second is what the cost of downtime was when you were expending that level of effort. If your organization is like the vast majority of industrial plants in the world, you will see decreasing downtime with increasing PM and PdM effort.

Now, plot a third curve: the addition of the first two. As you continue to add PM effort, you will reach a point where the first two curves cross and the third has a minimum. This is the economic point. Applying more effort to PM or PdM after this point makes no economic sense. You are actually costing the company more money by doing more.

Anytime that the cost of maintenance labor or the cost of downtime changes, re-plot the curves. This will keep you applying just the effort necessary to minimize costs.

Equipment Identification Scheme

All equipment is given an identifying code called the Equipment Code. This code is used to tie work orders, equipment files, history files, purchase orders, etc. to that unique piece of process equipment. This code does not identify a piece of equipment so much as a location in the process. A pump taken from stores has no code assigned to it until it is put in a particular function in your manufacturing process. Then, it assumes the code of the function it is filling. Other pieces of equipment such as fork trucks or large vessels will probably always have that one code assigned since they are rarely, if ever, changed.

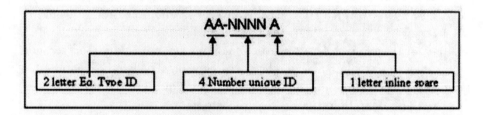

The Systemaitec code is a seven-digit alphanumeric symbol. It is made up of a two-letter equipment type designator. This can be found on the accompanying table for most common types of equipment. On the other hand, you may decide to make your own table based on your plant equipment. So long as you are consistent, it doesn't matter. The next four digits are numbers that uniquely identify the location. The last digit is a letter that is only used if the location contains an inline spare, such as a pump that is piped in but not used unless the other pump in that location fails.

Unique Numeric Identifier

This four-digit number is not a sequential number. Each digit can represent specific location data. Normally, a manufacturing plant is divided into specific areas, each having an account designator given to it for accounting purposes. For example, a sausage plant may have five areas: Raw Material Storage, Grinding and Mixing, Stuffing, Packaging, and Finished Cold Storage. Accounting may say that all expenditures for the Raw Material storage will be to accounts numbered from 1000 to 1999. All expenditures for Grinding and Mixing will be to accounts numbered from 2000 to 2999. And so on. So, they have already divided all the equipment, buildings, and services into a 1000 area, a 2000 area, 3000,

4000, and 5000 areas. Maintenance would be 6000. Distribution would be 7000. Admin would be 8000. Utilities 9000, etc. In addition, they may divide the particular areas up into sub-areas. Perhaps in Raw Material Storage, there is cold storage, ambient storage, and bulk liquid storage. Accounting may then say that all accounts in cold storage will start with 10; all accounts in ambient storage will start with 11, etc.

If your accounting department has divided the facility into major accounts, then the first digit of your unique identifier would be the first number of the account that covers that equipment. If your accounting department has subdivided the areas, then the second number is also defined for you.

The designator for a centrifugal pump in Bulk Liquid Storage would be PC12##X. The designator for a belt conveyor in Ambient Storage would be CB11## X.

If your accounting department has not set up those kinds of accounts throughout your facility, then you have to do it yourself. A word of caution: It is better to use the accounting numbers if they exist. Once a standard has been set, stick to it. It makes it much easier for people to learn.

Suppose no one had ever divided the sausage plant into accounting sections. Then you, the maintenance manager would do it. Designate Raw Material Storage as the 10 area, Grinding and Mixing as 20 area, etc. Set up subsections if that makes sense for your plant. Then use your area numbers as the first two digits of your unique equipment identifier.

The next two digits are arbitrarily assigned. You can assign them alphabetically, or according to process flow, etc. However you do it, be consistent. Start with 00 and work through that type of equipment in that area progressing one number at a time. If you end up with more than 100 centrifugal pumps or centrifuges or some other type of equipment, then you will have to go to five numbers or subdivide the area. Most manufacturing plants do not exceed this number.

There is a misnomer in what I call these four numbers. Although they are unique for any particular type of equipment—e.g., no two centrifugal pumps will have the same four digit number assigned to them—they may not be unique across equipment types, i.e., you could have PC-7010 and PD-7010 but you could not have two PC-7010's except in the case of inline spares (covered below.)

Inline spares are a special case. Because inline spares do the same job in the process, that is they both occupy the same location in the process, they both have the same number. The way they are identified uniquely is with the final digit. The final digit is used this way:

No inline spare	blank
Left spare	A
Right or middle spare	B
If a middle spare	
exists then Right spare	C

If you have more than three inline spares (very rare except on the Space Shuttle) then letter them sequentially from left to right when facing the rear of the equipment.

When you have identified all your equipment and given them equipment codes, then it is time to make these known to the world. Several things need to be done immediately and at least one other needs to be done in the near future.

First, stencil the equipment with the equipment code. Use large letters that stand out. The stencil should be applied to the base of the equipment—something that is not going to be changed out on a frequent basis. On a pump, the grouted base should be used. On a gearbox, the mount should be used. Every piece of equipment should be marked. It is much more efficient for the machine operator to simply glance at the subject of his/her work request to find that number than to look it up in a book or on a computer. Some people choose to use a brass plate. I find that these are normally too small to be of much use. The large stencils work much better.

Second, tables of common equipment names ("The Bulk Liquid Transfer Pump," for example) vs. equipment code (PC-1011) should be published and distributed in paper form or on the company network, if available.

Third, these same tables should be made available to the equipment operators where they normally fill out work orders.

In the long term, the equipment codes should be transferred to your P&ID's, PFD's, or other drawings and schematics normally used by operators, engineers, and maintenance technicians. This includes putting them on any computer screens if you are using distributive or computer control. If you are using control panels to operate several pieces of equipment, the equipment codes should be included in the switch and button labels on the panel.

If you already have equipment and/or history files, these should be re-filed by equipment code or cross-referenced to that code.

Equipment Type Designators

Equipment Type	Sub-Category	Designator
Automobiles/Trucks	Passenger	YP
	Light Trucks	YL
	Heavy Trucks	YR
Boilers	Gas fired	BG
	Oil fired	BD
	Waste Heat	BW
	Solar Powered	BS
	GeoThermal	BG
Buildings	Administrative	FA
	Process	FP
	Support	FS
Centrifuges	Vertical	GV
	Horizontal	GH
Compressors	Centrifugal	SC
	Piston	SP
	Other	SO
Conveyors	Air	CA
	Belt	CB
	Screw	CS
	Misc	CM
Dust Collectors/Air Cleaners	Tornadic	WT
	Filter	WF

Equipment Type Designators (Continued)

Equipment Type	Sub-Category	Designator
	Electrostatic	WE
Electric Motors	Single Phase	UP
	3 Phase <= 480 VAC	UT
	3 Phase > 480 VAC	UH
	DC	UD
Engines and Power Sources	Gasoline Engines	EG
	Diesel Engines	ED
	Natural Gas Engines	EN
	Gas Turbines	EA
	Steam Turbines	ES
	Water Turbines	EW
Flares	Ground	LG
	Tower	LO
Gear Boxes	Right Angle	DA
	Inline	DS
	Planetary	DD
	Gear Heads	DH
Generators	Motor Driven	QM
	Wind	QW
	Solar Panel	QS
	Motor-generators	QG
Heaters/AC	Water Heater	HW
	Air or Building	HA

Equipment Type Designators (Continued)

Equipment Type	Sub-Category	Designator
	Air Conditioner	HV
Heavy Equipment	Fork Trucks	YT
	Back Hoe	YH
	Crane	YC
	Front End Loader	YE
Machining/shop Eq.	Lathe	ML
	Mills	MM
	Grinders	MG
	Presses	MP
	Drill/Boring	MD
	Misc.	MI
	Shop cranes	MC
Pumps	Centrifugal	PC
	Positive Displacement	PD
	Diaphragm	PG
	Progressive Cavity	PP
	Misc.	PM
Reactors		RX
Refrigeration	Coolant based	FC
	Adsorption	FA
Relief Devices	Relief Valves	VV
	Conservation Vents	VN
	Explosion Vents	VX

Equipment Type Designators (Continued)

Equipment Type	Sub-Category	Designator
	Rupture Disks	VD
Unique Manufacturing Equipment		A
Vacuum Systems	Vacuum Jets	IJ
	Mechanical	IM
Vessels	Storage	TS
	Process	TP

Maintenance Reliability Systems

There are two systems that are basic to a good maintenance program. The first is a good operating system—getting work into the department, processing it, completing it, and notifying the customer of what was done. The second is this system: Reliability.

Some consultants and engineers would read that name and say to themselves, "Statistics." In many maintenance schemes, reliability is a very narrow statistical analysis of failure frequencies and redundancy schemes. Systemaitec takes a much broader view of reliability.

In this module, we will set up Preventive Maintenance (PM) and Predictive Maintenance (PdM) systems. PM systems are meant to increase the life of equipment. These systems insure that the equipment is lubricated, adjusted properly, and is not wearing at an excessive rate. PdM systems are meant to discover problems with the equipment before any signs are evident to the naked human senses. These systems insure that the rotating equipment remains balanced, that bearings are not being damaged, that shafts are aligned properly, that gears are meshing properly, that lubrication is not breaking down, that excessive internal wear is not occurring. In addition, with experience and good data, these systems will allow you to predict the life remaining in a piece of equipment that does have a problem so that you can schedule the repairs during non-peak production time saving time and money.

Preventive Maintenance

Lubrication

No piece of mechanical equipment can run without lubrication. Insuring that the lubrication is continuous, of the correct type and viscosity, and remains clean is the most important preventive maintenance that can be done.

Lubrication Survey

Systemaitec recommends that you do a lubrication survey. This involves a search of all vendor information for recommended lubricant types and viscosities. If no information is available from the vendor, then your lubricant supplier should be contacted and asked for a recommendation.

Choosing a lubricant is a complicated process that depends on ambient temperature, expected temperature rise, rotational speed of the equipment, weight of the moving member, cleanliness of service, metallurgy, seal material of construction, and many other subtle parameters. Assuming the machine is being used as designed, the OEM's recommendations are normally the best. If it is self-designed or of a modified design, you may need to contact a tribologist to make the call.

The survey not only lists the recommended lubricant but also predicts what the proper replenishment quantity is. Each lubrication point should have information about what to use, how much to use, and how often to replenish it. The lubrication survey should be kept evergreen as the plant ages and new equipment is installed. The lubrication survey provides the baseline for all the other lubrication programs.

Lubrication Route

Without lubrication, machines will fail within minutes. It is utterly important to keep all of the lubricants at proper levels. On the other hand, many bearings fail from being over-lubricated rather than under-lubricated. Too much grease or oil in a bearing housing will cause churning of the lubricant, heating it up, lowering its viscosity, causing wear on the bearing, causing more heat, causing even less viscosity, etc. Many organizations think that the right way to lubricate the machinery is to hang a grease gun near it, or leave a can of oil in the locker and tell the operator to lube it as it needs it or when he/she thinks of it.

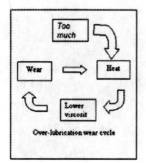

Over-lubrication wear cycle

This is a recipe for failure. Of course, it will never be diagnosed as a lubrication failure but as something else: the bearing failed; the gear wore out; the seal broke down. In reality, the failures all started with improper and inconsistent lubrication.

Systemaitec recommends that you devise a route that one or two people can use to inspect all the equipment. This route will let them look at every lube point in your process. Actually, we recommend three or four routes: a daily, a weekly, a monthly and a semiannually. Using the lubrication survey, the maintenance department should create forms for each route listing each lubrication point that needs to be checked at the route frequency, the type of lubricant being used, the replenishment volume to be used, and a place to record when lubricant was used and how much. We also recommend that each lube point be coded to match the route sheets. This can be with the machine name, etc but it is easier if one uses the machine code with a numeric addendum that designates that particular zerk fitting, lubricator, or oil fill cap. For example, on a small pump with a machine code of PC-7011 there could be a grease fitting on the rear bearing and an oil lubricator on the seal. These would be coded on the route sheet: PC-7011-L1 and PC-7011-L2.

Lubricator:		Monthly Lubrication Route Area 70		Date:	
Lube Point	Lubricant	Total Volume	Replenishment	Amount Used	
PC-7011-L2	Shell EP3	1 pint	When bulb is half full		
GB-7004-L1	Shell 90 wt	1.5 gall	When low in sightglass		
SC-7005-L1	Grease	--	One shot		
SC-7005-L2	Grease	--	One shot		
SC-7005-L3	Shell EP3	5 pints	When low on dipstick		
PP-7105-L3	Grease	--	One shot		
UT-7105-L1	Grease	--	One shot		
AZ-7101-L1	Grease				
AZ-7101-L2	Shell EP3				
AZ-7101					

Example of a Monthly Lubrication Route Sheet.

It is good practice to label the lube points on the machine. In very clean applications, the name of lubricant used should also be put on that label.

Who should do the lubrication? Many people feel the equipment operators should lubricate their own machines. Others feel that only maintenance should do that job. Systemaitec recommends a combination. Operators should do the daily and weekly routes. Maintenance should do the monthly and longer period routes. Why? Production is normally short term oriented. They often have several rotating shifts. Scheduling and monitoring a short cycle job is relatively easy for them. Scheduling and accomplishing long cycle jobs is much more problematic.

No matter who does the lubrication, they should only do it by the route sheet, should carry the route sheet with them, and should fill it out as they do it. The route sheet should be returned to maintenance when the route is complete. Maintenance must record all the lubrication use data in a database, either manual or computerized. This information is critical to maintaining and troubleshooting equipment and continuous improvement of the lubrication program.

Many companies are using dataloggers (small handheld computers that the lubricator carries along) in place of the route sheet. This has the advantage of always having the correct route because the datalogger downloads all the current information before each trip. It also has the advantage of making the recording of the completed information simpler by uploading it to a database in a few seconds rather that several minutes or hours of clerical work.

Dataloggers may be the way for your company to go. They have a high initial cost and they can be damaged readily in the field, but they provide significant timesavings and accuracy over the pen and paper route.

Along with dataloggers, many companies are using bar-coding to identify the lubrication point. In a relatively clean environment, this can be advantageous. In a dusty or dirty environment, it can take longer for the lubricator to get the bar code reader to work than to punch in the location data.

Machine Audit Program

Machine audits are not just a visual or PM inspection program. In many cases, they include inspections for wear, but they are much more comprehensive. Systemaitec only recommends machine audits for those manufacturing processes that are highly automated and require very tight manufacturing tolerances—or where the raw material varies enough to cause operating and adjustment problems. Manual processes that require operator intervention on each step are not good candidates for this program.

Date:	Audit of	Audit Team:	
Estimated time: 1 hour 30 minutes	**Weber**	Name:	
Estimated crafts		Name:	
1 I&E		Name:	
2 Mechanic			
Parameter	**Standard**	**Found**	**Adjusted**
Rotor Speed	1850 +/- 50 rpm	1862	
Screen Vacuum	20 mm Hg +/- 3 mm Hg	21.5 mmHg	
Air Flow at Center	15 scfm +/- 1 scfm	13 scfm	15 scfm
Air Flow at drive end	15 scfm +/- 1 scfm		
Air Flow at d...			
Tooth to blade clearance			

Example of a Machine Audit Form

Machines are audited at two times. First, when the process shows unacceptable variation. If the product parameters begin to drift toward one end of the tolerance, or in the language of Quality, the upper or lower control limit, then the machine is shut down and audited. Auditing consists of measuring, recording and adjusting all the variables within the machine. During this type of audit, the machine is set back to standard so that the product comes back into the middle of the control range.

This means that there must be a set of standards. Where do these come from? Initially, these came from engineering. During the life of the machine, though, these come from the second type of audit. When the machine is running exceptionally well; when the product is right in the middle of the control range and the variation is small; when the volume being produced is higher than normal; in other words, when the machine is running great, it is time to audit. This time, no settings are changed. All settings are measured and recorded. A comparison is made to standard. If there are real advantages to the new settings, a change is made to the standard.

During these audits, the machine is checked for wear. Unusual or severe wear is recorded on the audit sheet. Repairs may or may not be made at that time. The purpose of the audit is to take a snapshot of the equipment when it is producing at a particular rate or quality, and perhaps, to change it back to standard. Needed repairs due to wear can be scheduled later when the equipment is not needed to make production.

I recently set up a machine audit program for a bottling and packaging plant. The particular area of concern for them was the area that formed, packed, and sealed the cardboard boxes that contained the four one-gallon jugs containing their product. The plant had been in operation for five years and this had always been a problem area. The box blanks were not always the same dimensions. The insert blank also varied. The method for getting the equipment to run was trial-and-error—every time.

To solve this problem, I created audit sheets with every adjustment available to operators on the sheet and a set of parameters to measure on the box and insert blanks. We also set up a database of all of those readings so that the process engineer could examine the data as it was collected and determine what the proper settings were for the equipment vs. any particular set of raw material dimensions.

The most important part of this program was the realization that this was a long-term study that was probably not going to produce over-night miracles. We set up to measure weekly on a routine basis, and whenever the equipment was experiencing jams—prior to any adjustments being made—and when the equipment was running great.

This program will eventually determine all the proper standard settings for this machine and produce charts that will allow the operators to "dial in" the machine settings for any set of raw material parameters.

Balancing

Rotating equipment vibrates. There is no such thing as a perfectly balanced rotor. The purpose of balancing is to keep the vibration energy below the level that will hurt the bearings, the seals, and the rotor itself.

You do not need to do your own balancing. In fact, for a small plant, in-house balancing is probably an extravagant luxury that is losing money. There are many good balancing shops around that can handle everything from pump rotors to turbine rotors and do it efficiently. What you do need is a balancing specification for all your rotors that you include with all purchase orders. You also need a way to insure that the rotor is balanced to within that specification when you install it and stays that way as you use it.

The method for determining if your rotor is balanced and stays that way is covered in the PdM section. The specification is something you need to decide on.

I cover this in PM because requiring your rotors to be balanced when installed is like requiring the tires on your automobile to be balanced when they are installed. If you do that, you increase the life of your equipment. Certainly, you can drive your car with unbalanced tires. It is not recommended and it will cost you a lot of money in the long run but you could do it. Just so, you can install unbalanced rotors and run the equipment. It just doesn't make economic sense to do so.

What specification should you use? That depends on how the equipment is used; what speed it turns; what it is moving or what is moving it; and how much you are willing to spend for increased life. Normally, this specification is given in oz-inches. It measures how much unbalanced weight you will allow at a given radius. Tighter specifications cost more money but insure longer bearing, rotor and seal life. Looser specifications are cheaper but will result in shorter equipment life. A good specification for you cannot be determined without examining the complete system.

Written PM program

A walk through by a qualified mechanic can be very beneficial. It is not a PM program, though. To get the most from the time your people spend on walk-throughs, you need to send them on a specific route, tell them what to look for and record what they find. This is done with a written PM program.

CMMS normally includes a module on PM. There is a way to determine what you want to check on what frequency. The computer schedules it and generates a work order when it is due. The computer also records the findings and puts them in a database for future use.

For a small organization, a manual system can work just as well. It is more cumbersome and costly to run but has the big advantage of low initial cost. Normally, this would simply be a 52-week file. Each file folder would hold the work orders for tasks due that week. When scheduling the week, the planner or foreman would pull those orders from that file, copy them, and assign them. When they come back from the craftsperson, they are filed back into a parallel 52-week file, or recorded on a schedule sheet or board and filed in the equipment file.

What should you schedule? That can only be determined by examining your equipment, referring to vendor information and recommendations, and looking at your history file. This is a large initial task that has to be done before either a CMMS or manual system can be set up.

Instrument Loop Calibrations

Is your process automated? Almost all manufacturing processes are to some extent. That means that the information coming from your instrument loops is critical to speed and quality.

There are two ways to maintain an instrument loop: operator notification or loop calibrations. In the first, nothing is done to the loop unless the operator notices that the instrument is not operating properly. Then a work order is filled out and maintenance responds. The second is to routinely check and calibrate all loops on a scheduled basis.

If you have a distributive control system, I am sure you already do routine instrument calibrations. If you are ISO9000 or QS9000 certified, I am sure you already do those checks because they are required. If you are still running a board controlled process, you may or may not be doing this. Systemaitec recommends that everyone using instrument loops to control the process or feed information to a DCS do routine instrument calibrations.

With that, if you are using pneumatic controls, pickups, or sensors, then Systemaitec recommends that you do a routine Instrument Air dew point check at least once a quarter. It should never be higher than -40° F. If it is, then change your desiccant or troubleshoot your drier problems.

Shaft Alignment Kit

Some would question why I would put a piece of maintenance equipment in a section on PM. Properly aligned shafts will do more to increase bearing, seal, and rotor life than any other single thing you can do after lubrication. If you don't own a shaft alignment kit then you probably are not aligning your equipment properly.

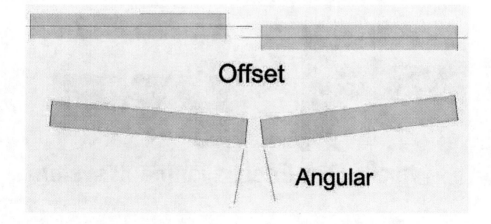

Most coupling manufacturers will tell you that their couplings can take up to a degree of misalignment. That is a pretty gross figure. They are correct, though.

They design couplings that will not wear out with that much misalignment. Unfortunately, the life of the coupling is not controlling here. Badly aligned shafts, and by that I mean much less than the one degree of misalignment the coupling manufacturers use, will ruin the bearings on the equipment in short order. All shafts, even low speed ones, must be aligned to within a few thousands of an inch TIR if the bearings are to last for their full expected life.

There are some tremendous systems on the market for alignment. They use lasers, computers, and proximity sensors. They will practically move the equipment and install the shims. They do no good at all, though, if they are not used by trained, qualified technicians who understand what the systems are telling them.

Systemaitec recommends that you have a working shaft alignment kit. It does not have to be a laser kit with computer control. A simple rim-face or reverse indicator system will do just as well for most applications. You can use a computer, a calculator, a pencil or a plotting board to compute the values of movement needed. The important thing is that your technicians understand how to do an alignment and the importance of doing so.

Systemaitec recommends that you set a standard for alignment based on TIR of the dial indicator or sensor. For most applications this should be no more than 0.001 inch per inch of shaft diameter. In some high-speed cases, it must be less than 0.001-inch period.

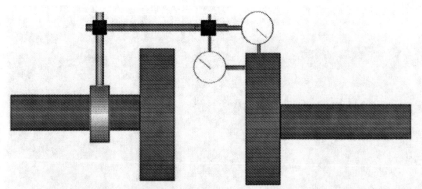

Typical Rim-Face alignment system

Equipment Files

This may seem like another strange topic for a section on PM systems. Maintaining up to date and complete equipment files, though, can be one of your best preventive maintenance techniques. Most vendors pack their manuals full of necessary information about the equipment. They make recommendations for maintenance, lubrication, and operating conditions. They tell you how to move and install the equipment. Having that information available to your crafts will keep your equipment running at peak and make repairs go much quicker. Equipment files, since they really only cost you a couple of filing cabinets and a minimum of clerical time, are one of the best bargains in maintenance.

In a new plant, early in my career, I ran across a metering pump I had never seen before. These pumps never seemed to work properly. They were not linear and, in some cases, would quit pumping in the middle of the process even though the mechanism was still moving. The mechanics complained about them because we were always pulling them, rebuilding them, and finding nothing wrong with them. A check of the equipment files showed that there was nothing there but the original brochure used by Engineering to purchase the equipment.

I finally called the vendor to come in and assist in a rebuild and troubleshoot the operation. The vendor's technician went through the whole rebuild shaking his head. "There's nothing wrong with this pump. Why did you pull it?' I explained the erratic operation and the failure to pump anything. His eyebrows rose, then he reached into his briefcase and pulled out a manual.

"Did you adjust the refill valve according to the directions?" he asked.

I grabbed the manual, read for a few seconds, and then with a red face, said no.

We reinstalled the unit, filled it with oil, started it up, spent a minute adjusting the refill valve according to the manual and walked away from it. It stayed in place for over a year with no more trouble.

Had someone put the manual into the file when the equipment was installed, we would have saved countless hours of downtime and an expensive visit from the manufacturer.

I was doing a maintenance audit a few years ago in a client's plant. One of the things I always ask for is to see the equipment files. In this particular plant, the files were not filed by equipment number, they were all located in one filing cabinet, and they seemed to be jumbled. My theory was that the crafts seldom used the files. I was insured by the engineer who was escorting me that this was untrue. His mechanics always used the files and the files were complete.

To test this, I asked for the brand of centrifugal pump most often used in the plant. The crafts responded quickly with a popular brand. I then asked, "How do you set the head clearance on that pump?"

There was a lot of shuffling of feet and averted eyes. I said to them, "Come on, guys, I don't expect you to know off the top of your head. Look it up."

It took them fifteen minutes to find the manual in the poorly sorted drawers, and another ten minutes to find the procedure for setting head clearance. I then asked, "So, when was the last time any of you set the head clearance on one of those pumps?"

There was no answer. They had been changing those pumps for years but not one of them had ever set the head clearance and did not even know that one was supposed to.

Yes, equipment files are definitely part of PM.

Predictive Maintenance

Since electric motors; gas, water and air turbines; and most other sources of energy turn, almost all manufacturing processes start with rotating equipment. There are two basic PdM systems for rotating equipment: Vibration Monitoring and bearing condition monitoring. Whether you need one, both or none depends on the size and speed of your rotating equipment and how reliable you desire your equipment to be.

If you only have very low speed (< 500 rpm), lightly loaded rotors, you probably do not need either of these systems. On the other hand, if you have very high-speed (>8000 rpm) equipment such as gas turbines or steam turbines, you definitely need vibration monitoring. For most manufacturing situations the speeds are somewhere between these two. Systemaitec recommends that you use a bearing condition monitoring system for any shaft under 1500 rpm and vibration monitoring for any shaft over 1500 rpm.

Vibration Monitoring.

All rotating equipment vibrates. The trick to good design and good maintenance is to keep the level of vibration below that which damages the equipment. Bearings are especially susceptible to damage from vibration but seals, both lip and mechanical, also suffer short lives because the shaft they are on vibrates excessively.

Vibration monitoring is not vibration analysis (VA.) Vibration monitoring lets you know when VA is needed. In many cases, though, the results of the monitoring itself can give you all the information you need to solve your problem.

Generally, every bearing on critical equipment is monitored for vibration once a month. On some particularly critical pieces or ones that have had a history of bearing failure, the monitoring may be done weekly. The frequency of vibration, the amplitude or vibration, and/or the velocity of the vibration are recorded. Over the life of a bearing these numbers show a predictable rise to failure. The advantage of this method is that sudden increases in vibration are directly tied to bearing or seal wear. The plant doing vibration monitoring is normally not caught unawares by a bearing failure. When the system is operating properly there are no more catastrophic bearing failures. All are predicted far enough in advance to repair long before failure occurs.

Monitoring requires the purchase of test equipment. Normally, this would be a hand held monitor with a data-recording device built in. The route is downloaded to the recording device. The operator walks the route, placing the probe on a specified point on the bearing housing. By pressing a few buttons on the device, the readings are stored and the next point on the route is referenced on the screen. At the end of the route, the operator uploads the data to a computer containing the route database.

Equipment Code			Equipment Name			Vibration point			
AI-7101			Fiberizer			Drive Bearing			
Week	10/5	10/12	10/19	10/26	11/2	11/9	11/16	11/23	11/30
Amp (mils)	2.1	1.9	1.8	2.1	2.2	2.25	2.7	3.0	3.5
Vel (in/sec)	.1	.11	.09	.15	.17	.2	.3	.55	.7
Severity	Sat								EX

Typical Vibration Monitoring Report showing a deteriorating Bearing

Reports are generated for each vibration point. They show the vibration levels over time. Most of the software identifies problem equipment and prints out a cursory analysis of the problem and recommended solutions.

Vibration monitoring alone should not be depended upon for low speed (<1200 rpm) shafts. Vibration monitoring equipment is unreliable at these low speeds.

Vibration Analysis

This program uses sophisticated electronic equipment to analyze all the frequencies and amplitudes that are in the vibrations of your equipment. Through long observation and analysis, most particular modes of vibration have been identified. A skilled technician can separate unbalance from bearing noise from shaft misalignment from blade pass frequencies from structural vibrations, etc. A highly skilled technician using good equipment and with a strong database of readings on a particular machine can effectively predict remaining life in most situations. This becomes a big advantage when attempting to schedule all maintenance during off-peak or no-production times.

In one case in which I was involved, a gearbox was failing on the agitator of a chemical reactor. Changing the box would involve about eight hours, even if a new box was ready to go in. The production department was behind on orders and could not afford any downtime. They wanted to know what the best route was: play it safe and take the equipment down for scheduled maintenance—losing the eight hours of production time and shorting some customers, or taking the chance on running the equipment and having it fail in the middle of the night—causing even more downtime as people were called in and heavy equipment was located on an emergency basis. After several hours of data analysis, the technician and I agreed that there was probably about 28 days of life in the box before a catastrophic failure would occur due to bearing and gear failure. The production department took our word for it. They got through their production crunch and met all their customers' needs. The box failed catastrophically in 26 days—well past the problem time for production.

Systemaitec does not recommend the purchase of vibration analysis equipment. The equipment is very expensive. The training is lengthy and the level of expertise needed for good predictions and analysis takes a long time to develop. There are many good firms (including us!) that specialize in vibration analysis. When needed, contracting those services is almost always cheaper than developing your own expertise.

Bearing Monitoring

On low speed rotors, vibration monitoring often misses bearing noise—because this noise is white in nature and very low in amplitude. It is necessary to use a different type of analysis to detect bearing damage on low speed equipment. It is, in fact, advantageous to use bearing monitoring on all equipment to insure the earliest detection of bearing wear.

One approach is to use SPM® or Shock Pulse Monitoring. This is an acoustical measuring tool that analyzes the noises coming from a bearing and looks for the particular type of noise that is generated from various wear modes. SPM® requires a sophisticated and expensive machine. Systemaitec does not recommend this approach for small to medium sized plants. There is a cheaper and more convenient way.

Most good vibration analyzers now use demodulation technology to identify white noise and the shock frequencies that are causing that noise. With the same machine that is identifying vibration, the maintenance department can identify bearing noise on any speed shaft.

Oil Analysis

Do you have large gearboxes or crankcases in your equipment? Do you change oil on a routine basis—every three months or some other frequency? You may be wasting a lot of money and you may be losing an opportunity for some great PdM.

Most lubrication distributors offer an oil analysis program. A sample is drawn from the oil reservoirs of the equipment on the program on a routine basis—sometimes once a month, more often once a quarter. The sample is sent to a lab where the condition of the lubricant is compared to a new sample. A written report is generated and returned. If the lubricant has no problems (i.e., it is not dirty, the viscosity is still in the proper range, there is no water in it, there has been no excessive oxidation, etc.) the recommendation will be to leave it in place for another cycle. In this way, you only change oil that needs changing. You save labor, the cost of the new oil, and the disposal costs on the old oil. This is a great deal for the cost of the analysis.

Oil Analysis Report

GR 7301
Blender Right Angle Drive Gear Box

Oil Viscosity: 130 cps – normal
Oil appearance: Clean
Oxidation: <1%

Metals:
Iron : 3 mg/l
Copper <1 mg/l

Recommendations:
Oil appears to be in good shape. Do not change this cycle.
The iron content is high indicating wear on gear faces. Since this is a new box continue t-

The real PdM advantage, though, is in what the lab can find in the oil. All gearboxes, crankcases, etc. wear to some extent. It is expected to find some wear metals in the oil. The lab analyzes for these metals. When an unexpected metal or amount of metal shows up, they notify you of the possibility of wear in in the box. Because bearings, gears, shafts, and seals are made of different materials, often the source of the metal can be identified, predicting what is going to fail. If you follow the recommendations of the lab and do not change oil out until it has to be, the increased wear metal in the oil can be tracked and a prediction made as to when failure will occur.

Systemaitec recommends that you install an oil analysis program for all equipment with oil reservoirs equal to or greater than two gallons and for production critical equipment with reservoirs greater than one gallon.

On Site Oil Analysis

There are new, hand-held machines on the market that will do a quick, brief analysis of the oil in your gearbox to determine if its properties are outside normal parameters. Although this machine will not give you a report about what is wrong with the oil, it will tell you if the oil is OK to leave in place for another cycle. In other words, it becomes a go/no go device for your oil sampling program.

How is this an advantage? It is estimated by the oil sampling labs that 80% of all oil samples result are nominal—in other words, 80% of the samples did not really need to be run. With the on site system, only those oils samples that need to be analyzed are sent to the lab—saving up to 80% of the money you are currently spending or would be spending on a conventional system.

These devices are expensive. If you have a small to medium sized plant you may not be able to justify the capital and training necessary to operate this system. There are, though, consultants (ahem!) that have access to these machines and can run the programs for you—still saving you money without the necessity of spending capital.

Thermographic Survey

This one you will like. Most industrial risk companies (insurance carriers) offer it for free. Thermographic Analysis is done with an infrared camera and an optical camera. A double photograph of the interiors of your energized motor control centers, breaker boxes, and substations are taken. When developed, hot points will be much brighter than ambient points in the infrared photograph. Comparison with the optic photograph identifies what contacts, fuses, breakers, or wires are generating excessive heat, indicating a short or bad connection. This is a simple process that, if done annually or semi-annually, will save you a lot of money and a lot of downtime. Most problems found with this system require a screwdriver or a new fuse to fix. Unrepaired, these can cost you major repairs of wiring and a lot of downtime.

Ultrasonic Thickness (UT)

UT uses an ultrasonic thickness-measuring device to check for wear in pipes, reactors, and storage vessels. Knowing that your pipes or vessels are corroding and at what rate they are doing so allows you to repair them on a scheduled basis.

To make this system useful, a series of points on the equipment or piping run must be precisely marked. On a long-cycle basis, readings are taken with the device at precisely the same location each time. The thickness measured is recorded in a database. The data is periodically analyzed to determine corrosion rate.

Systemaitec does not recommend this for most industries. The exceptions are the chemical industry, and those industries that transport abrasive slurries or corrosive chemicals in pipes.

If you are a PSM plant, this program is mandatory.

Maintenance Planning Module

Maintenance Planning!

That is a topic that elicits a lot of emotion—mostly because it was so abused in the 80's. Planning was going to be the way to double maintenance productivity. Every job had to be planned. All parts had to be staged. Craftsmen were viewed as machine that needed to be fed parts and information. What resulted in most places was bureaucracy that ate up resources and provided little benefit.

In reality, every job does need some level of planning. Every job and every crew, though, does not necessarily need the same level of planning. The secret to good planning is how much planning to apply to each job.

Although every job needs some level of planning, not every crew needs a planner. For crews of less than five, a planner will almost never increase efficiency enough to pay for his/her own position. If your crew of five is doing so poorly that a planner can increase their direct work by 20% each, then you probably have a discipline problem and planning will not do you any good until that is solved. For crews of between five and eight people, planning can at least break even by increasing the efficiency of the people in the field. The larger the crew, the more money can be made by effective planning.

In small crews, the level of planning can be very basic—time estimates, equipment needs, craft coordination—and be done by the foreman. Staging of parts and materials is not normally economical in a small plant maintained by a small crew.

In larger plants and crews, determining the most economical level of planning must be done on a plant-by-plant basis. The proximity of your shop and your stores to the plant equipment, the level of complexity of the tasks normally assigned to maintenance, the experience and qualifications of your craftspeople all play a role in how much planning is economical.

In most circumstances, where a planner is planning for just two or three crafts and only needs to identify the crafts, major tools, materials and provide general instructions, and has a good database of preplanned or previous performed jobs, one planner can typically handle up to 30 craftspeople. On the other hand, if it is a union shop with a variety of crafts and sub-crafts to plan and coordinate, and the planning needs to be very detailed, one planner may not be able to plan for more than eight to ten craftspeople.

The most basic level of planning that has to be done in any well-run shop is estimating the crafts and the man-hours for each work order that comes in. Every department does this. Some do it in a very cursory manner and keep no records. They also end up with nothing against which to judge performance of one crew vs. another. Still, the estimating gets done or the work would never get scheduled.

A good scheduling system involves good estimates. The planner or foreman needs to be familiar with the job to be done, to analyze how it is going to be done, and determine who and what needs to be at the job site when. There are many CMMS systems that will print a schedule for you once you have done this work. I don't know of any of them that are sophisticated enough to schedule without this upfront information being fed to them.

With a work order history file, computerized or manual, the planner has information available that keeps him/her from re-inventing the wheel each time. This history file can be a great time saver when planning repetitive jobs, even if they only repeat every year or so. Make sure you keep all the information that comes back to you on a completed work order. It is gold.

Scheduling is an art as much as a science. Since the planner is working with estimates, his tightly devised schedule is sure to dissolve into chaos after the first two minutes of the day. The planner needs to insure he has enough contingencies built in so that the foreman can adjust the schedule on the fly and still get all the crafts to that last job in the afternoon on time. Is this easy? Absolutely not. Most daily plans are no more than 80% effective. The trick is to build a knowledge base of past work that can keep the plan realistic and make it last further and further into the day.

Basic Planning Process

When an approved work order is received by the planner, he/she records the order in the inactive or unplanned log. According to priority, he/she retrieves those orders for planning. Each order is studied to insure all the information needed is there. If not, the planner contacts the originator or the authorizer to obtain the needed information.

The planner checks the history file to see if that job has been planned and performed in the past. If so, the planner extracts the information from the history file, modifying the original plan according to the comments in the completion section, and copies the new plan onto the order. If parts or materials that are not available on site are needed, he/she orders them and marks the order as "Awaiting Parts" (either in CMMS or in the manual system.) The order is then returned to the inactive backlog. If the order is ready to complete, it is moved to the active backlog. At that point, he/she is done with the planning.

If there is no information in the history file, then the planner goes to the job site. He/she lists the special tools that may be needed, the special equipment (fork lift, crane, jack stands, etc.) that may be needed, the parts or repair kits that will be needed, determines the type and number of crafts and the order of their work, then estimates the clock hours necessary for each craft. Back at a computer, the planner determines the stores numbers of the parts and lists that with the other information. With that information in hand, the planner writes up a planning document and attaches it to the work order. The work order is then placed in the active backlog.

If the plan calls for parts or materials that are not in stores or on site, the planner orders the parts or materials, marks the order as "Waiting Parts" (either in CMMS or in his/her manual system) and returns the work order to the inactive backlog.

When the parts or material arrive for a work order, they are marked with the work order number and put in a staging area. The order is retrieved, the note removed, and the order put in the active backlog.

At the end of each day, the planner or the foreman, in consultation with a representative of the production department, picks enough work out of the active backlog to completely fill the time of the crew available. This starts with PM (Preventive Maintenance) and PdM (Predictive Maintenance) orders that are due, then moves down through the priority and aging system.

The morning of the plan, the foreman meets with the production department to insure there has been no change in priority or that there are no emergency orders to schedule. If there are changes, the foreman makes the changes in the schedule as necessary, assigns the work to each craftsperson, and notifies the planner of the change in schedule.

In the afternoon, the uncompleted work is taken to the planning meeting. At that point a decision is made to continue the work the next day, hold craftspeople over to complete it that day, or to return it to the backlog to await future scheduling.

Daily Schedule Sheet
Finishing Department Crew
Date: _____

Typical Daily Schedule sheet

W/O#	Equip	What	Craft	Name	Time	Notes
A00BC2	PC3201	Change seal	M	Jones	0800	
A00AZ1	TS1010	Open for inspections	M	Prior	0800	Leadman
A00AZ1	TS1010	Open for inspections	M	Reily	0800	Prior has W/O
A00AZA	CB3001	Stiff support roller	M	Gregory	0800	Need to lock out
A00AZA	CB3001	Break motor coupling	E	Jessup	0800	Done first thing.
A00BAA	PC3205	Check out noisy	M	Jones	1100	

For small crews (<5 craftspeople) with only one or two crafts, Systemaitec recommends this level of planning. The daily schedule should be a simple printed list of all the work to be done, who is assigned to do it, when it is scheduled, and special notes about craft interaction, if necessary.

Resource Loaded Scheduling Method

For larger work periods, such as a turnaround or maintenance shutdown, we suggest you use a Gantt or PERT or Critical Path chart. For a small crew, though, we suggest another bar-chart method: Resource Loaded Charting.

RLC charts the resources you have instead of the jobs. It can be reversed and turned into a typical job loaded bar chart for use during the outage. The main advantage of Resource Loaded Charting is rapid planning of a specific list of work within the shortest time possible, and the identification of the need for additional resources.

All of us have heard the lament from the crews, "You guys plan like: 'if it takes one woman nine months to make a baby then nine women could do it in a month.'" Many times the lament has a good basis in fact. Using man-hour estimates to schedule work sometimes does leave us in the position of applying more resources than can be effectively used on a job in the mistaken belief that that will get the job done more quickly. In reality, too many resources probably slow the job down.

RLC is done as follows:

Craft	0800	0900	1000	1100	1200	1300	1400
Mechanic #1							
Mechanic #2		A00A01				A00AA2	
Mechanic #3							
I&E #2							
I&E #3	AA1		A00A23				

Work order #	Job Description	Craft	People	Hours
A00A01	Change Fiberizer Rotor	M	3	6
A00A23	Clean MCC	E	1	2
A00AA1	Remove motor on oven drive	E	2	1
A00AA2	Replace bearing on drive side of oven tail roll	M	2	4

This is best done electronically or on a magnetic planning board. We will look at how to do it using a magnetic board with flexible magnetic strips.

A bar chart is formed with the hours or days plotted horizontally. Magnetic strips are cut for each time interval desired. When I used this system I kept a drawer full of strips cut to half-hour lengths all the way from ½ hour to the length of the outage.

Down the left side of the chart is listed the craft resources available. If there are 10 mechanics and 3 I&E technicians, then Mechanic #1, Mechanic #2,..., Tech #3 are used as the labels for each bar on the chart. The list of work that must be performed during the period of time is listed. Instead of the man-hours, the planner puts in two separate columns the number of people that are to be assigned and the number of clock hours it will take to accomplish the job with those people. Labels are written with this information for each job and craftsperson (i.e., if the job is estimated to need three mechanics, then three labels are produced with that job number on them.) These labels are attached to a magnetic strip of the proper length to represent the estimated clock hours the job should take. All job strips are stacked together.

Starting with the first hour of the shutdown and at the top of the resource list, the planner begins placing jobs. If the job has three strips (i.e., it requires three of that craft) then the planner makes a bar for the first three craft resources with the three strips. He proceeds to put the jobs on the chart according to what resources are left. When all the jobs are on the board, the planner begins to "square up" the time frame. If all the jobs don't end together, he/she begins to manipulate the schedule to fill in blanks and to add resources to make the shutdown square and get it done in the smallest number of clock hours possible.

This method will insure the best use of all resources and the shortest shutdown possible. After the planning phase, it can be recopied as a standard bar chart with the jobs on the left and people assigned to the jobs written on the bars or in the resources column.

Higher Level Planning

For day-to-day work, higher level planning is not needed in most manufacturing plants. For some, though, what would be a luxury is a necessity. This happens when you have many crews with a lot of work rule restrictions in a union environment. Planning, then, may have to be very intense to get your direct work even near 50% (See the MES module.)

CPM Charting

In some instances Critical Path charting of each day's work may be needed to improve efficiency. This would be especially true in a large shop with many crafts involved, and with work that requires the interaction of many crafts on one job. The need to have only an electrician disconnect a motor coupling, and only an insulator remove fiberglass insulation, and only a boiler maker remove a cover from a tank, and only a machinist change a seal, and...You get the picture. Coordinating all those crafts efficiently is almost impossible. It certainly ranks up there with trying to get a space shuttle launched. That is when a higher-level planning technique like CPM or PERT becomes a necessity.

Many large organizations have solved this problem by accepting that day to day maintenance is going to be inefficient, that craftspeople are going to be sitting on their butts waiting for some part, other craft, or supervisor to show up and straighten out the mess. I do not recommend this approach. That sloppy attitude soon extends into every phase of maintenance. The result is normally a poorly running plant with very high costs and a very poor safety record. It would pay those organizations to hire one planner for every 8 crew members and produce detailed plans every day—then enforce them.

Parts Staging

In many instances, and especially in manufacturing areas that are large, spread out, or very remote from the maintenance shop or stores area, parts staging may be necessary to decrease travel time and increase direct work. During the planning process, the planner either procures all the parts necessary to accomplish the work order or notifies the storeroom of the need for them. Stores picks the parts and puts them on a pallet. In the most elaborate of the systems, the parts are delivered right to the job site at the same time the craftspeople are assigned to be there. In less elaborate schemes the parts are delivered to the maintenance shop the day before and staged on a pallet with the work order number stenciled or written on every box. The planner adds whatever materials he/she may have ordered for the job to the pallet.

In most cases, parts staging does not live up to its promise of eliminating trips back to the shop or to stores. Invariably there are things needed that the planner could not anticipate and the craftspeople have to return for a tool, for a piece of rope, for a tube of glue or something. It takes just as long to travel back to the shop for those items as it does to come back to get a whole pump.

Still, significant timesavings can be attained on jobs that are repetitive and routine—jobs where all the parts needed are assembled into a kit and kept together. If you have a travel time problem within your department (see Work Sampling in the MES module) then you should consider parts staging.

Detailed Work Instructions

Some systems require the planner to write out detailed step-by-step work instructions. They also require the craftspeople to follow them. Most of the time it is waste of everyone's time. You should be hiring craftspeople that know how to do the job. If the job entails changing a mechanical seal on a pump shaft, then that is all the instruction the craftsperson needs. Initial and recurring training takes care of the how-to (See the MTS module.)

There are two times when detailed instructions should be available with the work order:

- If there is a hazard associated with the job that is not apparent or out of the ordinary. These hazards are determined when a Maintenance Hazard Analysis is done for the job. We cover MHA's in the MSS module.

- If the job involves equipment that is unique to the plant and the task is not something that would be routinely taught in craft training. For example, I worked in a chemical plant that used a series of heavy rotors to slice a large block of polymer into small chips. The rotors weighed close to a ton and were covered with knife blades that were razor sharp. They rotated at about 900 rpm. The blades passed stationary bed knives twice per revolution. The gap between the rotating blades and the bed knives was 0.005 in. +/- .001 in. Failure to set these correctly could end up destroying the machine or ruining the process. The danger of losing a finger or even a hand was extreme every time the adjustment process took place. Although the job was done every few months, detailed instructions for doing this job with the safe work procedure accompanied every work order.

MAINTENANCE EFFICIENCY MODULE

In this module, we will establish metrics by which we will measure maintenance efficiency on an ongoing basis. Maintenance efficiency is not necessarily just dollars or hours or uptime, although all those things are included. Those are mostly lagging indicators. The real value of metrics is having leading indicators—those things that tell you what your costs, hours, or uptime are going to be in the future. We will establish a series of maintenance metrics that will be both leading and lagging—so we can see where we are going and where we have been.

What metrics will you use? In general, that cannot be determined prior to a detailed look at your organization, your facility, and your circumstances. There are a few metrics, though, that make sense for almost all maintenance organizations. We will look at those first.

COMMON METRICS

Systemaitec recommends that you establish these metrics immediately and begin to track them on a quarterly and annual basis. Graphs should be produced that track your progress against each of them. You, as management, should let your organization know that you are tracking them and let them know the results each quarter. Set goals for each metric and measure your progress toward them—publicly.

Costs

Maintenance Cost as a percent of replacement value (MTRV.)

This ratio is normally readily available from accounting. If the replacement value of your facility is not known, then have your Controller calculate it. There are GAAP rules for doing this that will make it a reliable number. Each year, the Controller or accountant should recalculate that value taking into account inflation and capital equipment that has been added.

How one defines maintenance cost is arbitrary. I would suggest you let your accountant decide what number to use. Normally, all costs are included. This means labor, parts and materials that are charged to maintenance accounts. If you use a charge-out rate for maintenance labor, the maintenance department costs are already included. If you do not use a charge out rate but direct charge labor to the production department, then you must add in any maintenance department charges.

If you include all costs, this ratio should be between 1.5% and 5%, depending on your industry. If you are outside of this range, then insure you have good numbers. If the numbers check out, then you have some real work.

Maintenance costs under 1.5% indicate that you are not maintaining your facility and are losing a bunch of money to breakdowns and deterioration. This is short-term gain and long term disaster.

Maintenance costs over 5% indicate that you are spending too much on maintenance. It could indicate too much PM or inefficient use of maintenance resources. Only a detailed analysis of the whole function will decide which it is.

Excellent organizations in low impact, low abrasion industries run between 1.7% and 2.5%. High corrosion industries, like certain chemical operations, can also run in the 2% range. High impact, high abrasion, and very high corrosion industries will typically run in the 4% range. With a well-managed maintenance effort, most manufacturing facilities can get their maintenance costs under 3.5%.

Maintenance as a % of total spending

Some people use this metric. I do not recommend it except for very static industries that have very consistent costs year after year. If the costs only vary by inflation and the depreciation, then this may be valid. On the other hand, if the plant is adding capital equipment inconsistently, or not at all, so that depreciation is constantly varying or always decreasing, this ratio will vary or increase with no change in maintenance. It becomes a number without a reason. No conclusions can be drawn from a change in the ratio without a lot of other data at hand.

True Cost of Maintenance Index

Many companies look at their Maintenance Spending to Replacement Value ratio and feel they are doing an excellent job because it is so low. As I stated at the beginning of this module, a ratio under 0.015 probably means you are spending too little on maintenance.

How can one tell if that is true? Is there some way to know if maintenance spending is too low? There is another index—developed by Systemaitec—that can give you that answer. Over time, it can also tell you what the most economic level of maintenance is.

There are actually two portions of maintenance costs to the operating plant: The cost of maintenance that includes equipment, labor, materials and overhead and the cost of equipment downtime due to maintenance. Normally, we do not calculate in the cost of downtime but it is a true cost of maintenance. It is this cost that will begin to rise when the replacement value ratio gets too small.

The metric is calculated as follows:

$$TCM = 10*(MTRV+MDT/SH*OROI*10)$$

Where:
TCM = True Cost of Maintenance index
MTRV maintenance cost to replacement value ratio
MDT = Hours of maintenance caused downtime.
SH = Scheduled hours of operation
OROI = Operating Return on Investment

Essentially, this index considers the amount of money that maintenance is costing compared to what it would cost to replace the whole facility plus the amount of return that is lost due to equipment downtime.

A very good facility with an OROI of 15%, 2.5% maintenance downtime, and 3% MTRV would end up with

$$TCM = 10(0.03 + 0.025*0.15*10) = 0.675$$

A facility with an OROI of 15% and a "world class" MTRV causing excessive downtime would have an index like this:

TCM= 10*(0.01+0.08*0.15*10)=1.3

So the TCM index would be twice as high even though they were spending at a "World Class" maintenance/replacement ratio.

A plot of TCM vs. MRV would show a curve like the following:

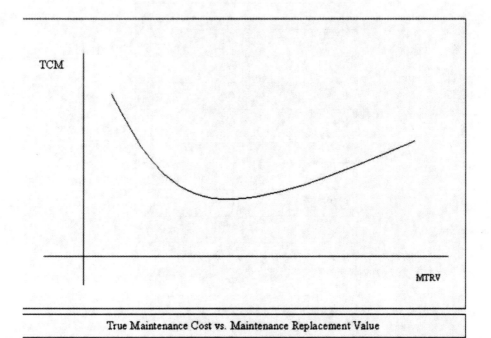

True Maintenance Cost vs. Maintenance Replacement Value

There will be a definite minimum in the curve. This minimum would define the proper economic level of maintenance effort. This curve should be calculated if the data is available. Firms just implementing the Seven Modules of Systematic Maintenance should collect this data and keep a running plot of this curve. The slope will tell you what side of the minimum you are on.

What is world class for TCM? We don't know. This index is so dependent on your facility's OROI (remember, it is your facility's OROI, not the overall company's return) that it becomes almost unique. We can say that for a facility with an OROI of 20%, a world class TCM would be 0.6. Below is chart of world-class indexes vs. OROI from 8% to 22%

OROI	TCM
0.08	0.36
0.09	0.38
0.1	0.4
0.11	0.42
0.12	0.44
0.13	0.46
0.14	0.48
0.15	0.5
0.16	0.52
0.17	0.54
0.18	0.56
0.19	0.58
0.2	0.6
0.21	0.62
0.22	0.64
0.23	0.66

Where do you fit on the chart? What does this number mean to you? It is probably more useful to look at your own history. The value of this index is in its ability to pinpoint the economic level of maintenance.

Labor Efficiency

Work Sampling

Work sampling is a statistical method for determining how much of the time for which you pay your employees is used for actually working on equipment. Using a statistical sampling technique, the surveyor monitors the crew's activities over a specified period of time. The samples are just that. The surveyor walks a prescribed path at random times and records what he/she sees each craftsperson doing when first observed. The activity observed is recorded into one of five categories: Direct work, Indirect work, Travel, Delay, or Non-productive. Direct work is recorded when the craftsperson is actively doing the assigned work, i.e., is at the job site and is actively engaged in the job. Indirect work is when the craftsperson is doing support work that will enable him/her or another craftsperson to do the job; e.g., they are reading drawings, arranging tools, setting up test equipment, etc. Travel is recorded when the surveyor sees the craftsperson walking, riding, or being transported to or from the job site or some other location. Delay is when the craftsperson is waiting for something beyond his/her control, e.g., waiting on parts at the storeroom or tools at the tool room or waiting for a work permit at the job site. Non-productive time is when the craftsperson is on break or standing or sitting around with no apparent job related activity going on.

Work sampling never records the name of the individual. It would be almost impossible to get enough samples on any particular person to make the statistical analysis meaningful. In fact, it is very difficult to get enough data on a small crew. When the crew is less than five, work sampling is probably not a viable way to gage efficiency.

Excellent organizations will have direct work percentages in the 80-percent range. Typical maintenance departments will be in the 50-percent range. Very loosely run organizations with no planning functions will be in the 10 to 30-percent range.

Repeat Failure Frequency

Many organizations use this metric. Anytime a repaired piece of equipment fails within a specified period, it is reported. This can be reported as a percentage, i.e., number of repeat failures per 100 repairs. This should be an extremely low number. In an excellent organization, it could be expected to be between 0 and 2. Normally, a failure analysis will be performed at the same time. If it was the repair that failed, an investigation is started to determine why the repair was not successful.

Work Sampling tells you how active your crew is in doing what was on the work order. Repeat Failure Frequency tells you how effective that activity is. They are both needed.

Backlog

Almost all organizations use some form of this metric. Backlog is the amount of maintenance work that has been ordered by the operating departments plus the amount of PM and PdM that is scheduled. Backlog cannot be accurately determined if there is no work order system in place. It also requires some rudimentary planning to be done—at least the estimate of how many man-hours of each craft is needed to accomplish the ordered work.

In CMMS systems, backlog is determined by the push of a key on the keyboard. In manual systems, it normally requires a review of the active and inactive work order files with a manual count of estimated man-hours.

As a metric for maintenance performance, backlogs should be measured in crew-weeks instead of man-hours. They should be plotted weekly. A healthy backlog is necessary to insure that the planning and scheduling process can function properly. If your work order system is working properly, and the operating departments are not getting serviced by verbal request, and your backlog is very low or continually dropping, you probably have a crew that is too large. On the other hand, if work sampling and repeat failure frequency are telling you that your crew is fully active and effective, then a low or dropping backlog indicates that your work order system is not functioning properly. You may not be getting all the work recorded, or your planner may be underestimating the labor involved in the jobs.

A well-run department should have a backlog of one to two crew-weeks. In essence, though, the department really wants a backlog that moves around some selected point, dropping when things are going well, rising when problems occur, when vacation time rolls around, or when there is a bunch of holidays in a row. The real leading value of this metric is in telling you how much labor you should budget for your next cycle. A constantly rising backlog along with satisfactory work sampling results indicate your crew size is too small and should be increased or supplemented with contractors. A falling backlog indicates that you have solved some of your maintenance problems and you probably need to increase your labor assigned to PM (if you have not passed the economic point) or decrease the size of your crew.

Work Request Aging Days

Sometimes, a backlog number doesn't tell the whole story about how your work is moving through the department. Aging days can give you an indication of what work is being assigned and what work is dragging out. Many times, people will work on the interesting or "important" work and try to delay the more mundane tasks. PM has a tendency to be viewed as mundane by most craftspeople. Insure that PM work orders are not aging. Other work, like cleanup and fix-up will sometimes age in the backlog. You, as the manager, must insure that this work is getting done in a timely fashion. It is, after all, work that has been ordered and authorized by your customers.

Aging days are calculated by averaging the total time that all work orders of similar priority have been in the backlog.

Systemaitec recommends producing an aging days report by priority range monthly. Below is a chart correlating the Systemaitec recommended priority system, the 4 priority system and recommended aging days for each.

Systemaitec Priority	4 Priority	Aging Days Max
90	E	0
70–89	U	7 days
40–69	R	14 days
20–39	R	54 days
2–19	R	90 days

In addition, you should always look at the oldest work orders in the system to insure that they are turning over. If a job has not been done in 90 days and no one has requested information on why, it is a good indication that it was not wanted in the first place. All work orders over 90 days old that have not been scheduled and are not waiting on parts or materials should be taken to the planning meeting and discussed.

EFFECTIVENESS

Downtime

The maintenance department exists, primarily, to keep the equipment producing a marketable product. We measure how well we are doing this with downtime. This number is calculated and tracked by the production department. Normally, there are two reasons for downtime: Maintenance and Process. It is up to the production department to code all the downtime with one of these reasons. In addition, once maintenance has installed the equipment code scheme, production should also be recording what equipment caused the downtime—especially for but not limited to maintenance downtime.

Maintenance normally can only affect and should only use the maintenance downtime number. This should be reported and graphed monthly. A statistical approach should be taken in measuring changes in this. Valid data analysis is necessary to insure that real information is coming from the data. I remember early in my career measuring maintenance downtime monthly and analyzing the data semiannually. My conclusions jumped all over the place. One period downtime was up; the next it was down. I did excellent regression on the data; produced trend charts and predictions based on the analysis; and was always wrong. When I finally went back and analyzed two years worth of data (24 data points) I discovered that nothing at all was happening. The regressions went almost flat—a slope near zero. I simply had not been analyzing a statistically significant sample.

Analyze downtime all you want. Don't draw any conclusion or make any promises until you have at least 25 data points. Until that time, use the supporting data to identify your problem machines and go to work on them. Then, when you have the 25 points, you can see if you are doing any good.

Percent Emergency vs. Routine

The reason you spend money on PdM is to decrease downtime. That means that, as you increase your reliability, you should decrease the amount of emergency work you do and increase the amount of routine priority work. Reporting this ratio over time tells you if your PdM is being effective. This ratio should be calculated two ways: the number of orders in each priority and the number of man-hours in each priority. It should not include backlogged work. Only work actually accomplished during the period should be used.

In a perfect world, the ratio would be zero. In reality, this ratio should be less than 0.1. While implementing your PM programs you should track this ratio in both forms to insure it is decreasing. If it is not, it is an indicator that your PdM is ineffective and you need to change how it is done, or how the results are interpreted, or how much importance the results are being given by both the maintenance department and the production department.

OTHER METRICS

There are other metrics you can use. For a small organization these may or may not give you useful information. Larger organizations might find them very useful.

MTBF

This is Mean Time Between Failures. This is a statistical approach that is used to predict how long a particular type of equipment should last after a repair. It would be very useful if you used a large number of a particular type of commercial machinery—such as a mass-produced centrifugal pump. Because it is statistical, it has to have a valid sample size and population or the data can be very misleading. I do not recommend this for small organizations. It takes too long to get enough data to be useful.

You calculate MTBF by researching the dates between failures of a piece of equipment and using a statistical approach to calculate the mean. Then, you track whether this is getting better or worse. When you do have many of the same model of equipment, MTBF can give you good information on how effective your repairs are for equipment that has occupied particular locations in the process. If the data collection is set up to also track the equipment as it moves from production to rebuild and back into production, MTBF can also highlight particular pieces of equipment that have problems.

At one point in my career, I was the maintenance superintendent for a PVC plant. The PVC reactors all had bottom entering agitators with mechanical seals. These seals were a constant problem. In a train of four reactors, one of them always seemed to be down for a seal change. It was not until we began tracking each seal cartridge through the rebuild shop and back into production that we noticed that one particular cartridge had a MTBF much smaller—almost half—of the others in the system. The next time it was out, we did a detailed analysis and found that most of the parts in the cartridge were near one end of the dimensional tolerance or the other. We scrapped that cartridge and solved our problem.

PM/PdM to total work ratio

This ratio tells you if the amount of PM you are doing is increasing as a percentage of all the work being performed. This is an interesting number but not of much use for diagnosis since it depends on too many factors. If you have calculated the economic point, you may want to use this number on a month-to-month basis as a leading indicator of when you need to recalculate that point.

Maintenance Stores Turnovers

This ratio is obviously only useful if you have a maintenance stores system. Maintenance inventory is very expensive to hold. It costs about 25% of the cost of the item annually to hold it in stores. Obviously, if it is not being used, you may want to consider selling it back to the vendor or scrapping it.

Maintenance Stores Stockouts

The only reason to pay the carrying charges on maintenance stores items is to decrease the amount of time maintenance has to wait on parts to repair an item. Stockouts are failures of stores to supply the parts when maintenance needed them. A stockout may indicate the need to increase your minimum quantity in stores.

I suggest that you track four different categories of stockouts—for they all have different causes and corrective actions.

The first type is when a demand is made on a part, the inventory systems says there is a non-zero quantity, the quantity is, indeed, on the shelf but the demand is higher than the quantity available and higher than the maximum allowed. This type of stockout indicates that your max may be too small.

The second type is when a demand is made on a part, the inventory system says there is a non-zero quantity, the quantity on hand is on the shelf but is below max, the demand is greater than the quantity on hand but lower than the max. This indicates that your order point may be too low.

The third type is when a demand is made on the part, the inventory systems says there is quantity on hand large enough to meet the demand but the actual number of parts on the shelf is less than the reported quantity and less than the demand. This indicates a problem with your inventory system and cycle counting.

The fourth type is when a demand is made on a part, the inventory system says there is zero quantity and there is, in fact, no parts available to meet the demand. Although this appears to be a special case of the second type, it is unique enough to deserve its own category. This type of stockout indicates that your mins are too low or your order point is too low depending on whether the stockout occurred because of a delay in delivery or an increased demand on the part.

Average time to complete tasks

This is a good statistic to use when evaluating the productivity of your crew. With a CMMS system in place it is relatively simple to obtain this information and keep it updated. Using a manual MMS system makes obtaining and maintaining this number very tedious. Whether it is useful to you depends on a lot of factors, not the least of which is what your other effectiveness numbers are telling you. If there is a problem—this number can pinpoint what and who. I would suggest that you report this number for those tasks that are routine and occur frequently. Report it by craftsperson, if you can, to see if there is a training or motivation need developing.

Failure Analysis

This can be a tedious metric but a highly effective one for some equipment. I would recommend that this be done for every mechanical seal and bearing failure. Other than that, it may not be useful for the small organization that does not have a lot of breakdowns.

Mechanical seals should not fail until the soft face has worn away. I don't know of too many that have lasted that long, but there is no real reason for them to fail in any other mode if they are installed and operated properly. Unfortunately, most seals are not installed properly and are not operated properly. Failure analysis lets you pinpoint the failure mode and take corrective action. There is a truism in life and it is especially true in maintenance. "If you want what you've always gotten continue to do what you've always done." Simply replacing the seal without changing anything is a guarantee that you will be changing it again, soon.

Bearings are in the same category. A properly installed and maintained bearing should last ten to thirty years. If it is not doing so, then failure analysis allows you to find out what went wrong and correct it. If you just replace the bearing, you will be replacing it again, soon.

Equipment Repair History Review

If you have the engineering resources, this metric will produce a lot of surprises and improvements. If you don't have those resources and can't afford to contract them, then this is not a metric you want to use.

All critical equipment history files should be reviewed annually for repeat failures, incomplete repairs, high cost components, etc. Then, you can apply resources to improve the reliability and decrease the cost of maintaining that equipment.

Summary of Recommended Maintenance Metrics

Metric	Best	Avg.	Poor
Maintenance Cost as a percent of replacement value.	<2%	3.5%	>5%
Maintenance as a % of total spending	—	~10%	—
True Cost of Maintenance Index	0.6	0.8	>1
Work Sampling Direct Work	80%	50%	<30%
Repeat Failure Frequency	0	<2%	>5%
Backlog		2 cr-wks	
Work Request Aging days P. 70–89		7 days	
Work Request Aging days P. 40–69		14 days	
Work Request Aging days P. 20–39		54 days	
Work Request Aging days P. 2–19		90 days	
Maintenance Downtime	0–2%	5%	>7%
Percent E vs. Routine	<10%	15%	>20%

Maintenance Training Module

The best maintenance systems in the world will not improve your maintenance efficiency if the craftspeople are not qualified. We look at two areas in this module: Hiring and Development.

Hiring and Evaluation

Ask yourself how you hire or evaluate an existing maintenance technician. In many areas of the country, finding a qualified, experienced craftsperson is getting very difficult. Many companies have had to resort to hiring young people who have an interest and aptitude and using them as their crews.

How do you tell if that person sitting across the table from you is qualified or not? You can ask questions. That certainly would get you their opinion of themselves. You can have other qualified craftspeople interview them—and take a chance on being sued for an EEO violation. If you only use an interview to evaluate them, you always take the chance of hiring someone who talks a good game but cannot perform.

Job Descriptions

The first step in hiring good craftspeople is to develop a complete job description. This isn't a list of jobs or skills that you expect them to have. It would be good if everyone who came through the door already knew how to adjust your compressor. That isn't very realistic, though. The job description is a list of abilities, physical, mental, and perhaps societal, that you expect of all your craftspeople in a particular trade. The key here is the word ALL. If you do not require your existing employees to have those abilities, don't put it on your hiring requirements. You can get in big trouble with the EEOC that way.

Divide your job description into three parts: Physical Abilities, Mental Abilities, and Societal Abilities.

Physical Abilities

If your crafts are routinely called upon to lift 50 pounds, put that in there. If they are routinely required to climb ladders at least 20 ft. tall—it goes in there. If they have to routinely walk over rough ground, or must routinely run a quarter mile in less than 2 minutes—whatever the job routinely calls for—put it in there. Just make sure you really do require everyone in that craft to be able to do those things because it is demanded by the job. Putting in requirements that are not actually part of the job will get you in trouble with the EEOC.

If they have to have certain aptitudes, put it in there. A mechanic should be dexterous enough to use tools efficiently. They should have a range of motion with their arms and upper body that allows them to turn large wrenches. They should be able to start small screws in threaded holes. They should be able to read a dial indicator or a rule to the smallest increment.

This part of the job description requires a detailed review of what the craft is required to do. Go watch your crafts work for a while before writing this. When done, check with your HR representative to insure it meets all the requirements of the law.

Mental Abilities

I have seen many job descriptions that say an eighth grade reading level is needed. Not only is this ambiguous (not every school's eighth graders read at the same level,) but it may also be illegal. What you really need in your job description is a list of things the person must be able to do. I emphasize must. If you have people currently working for you that cannot read, and you have not made an attempt to correct that, then you cannot make the ability to read a condition of being hired.

If you really mean that the mechanic you hire must be able to read a technical manual and understand what he/she has read, then say that. If it is necessary for a person in your maintenance department to read a dial indicator and estimate to a tenth of the smallest increment, then say that. If the person routinely has to add a column of decimal numbers together, make that the requirement. If the person is routinely called upon to add fractions, then say that they must be able to add fractions. If they have to multiply or divide or take square roots, then make those the requirements rather than saying, "must understand basic arithmetic."

Societal Abilities

Many companies are moving toward team based management systems. In those types of management schemes, the social skills of the candidate become important. "Plays well with others" is probably not a good thing to put in the job description but one can say things like: "has the ability to interact positively with others in a team setting." "Is willing to take responsibility and lead others." "Controls own emotions and handles emotional outbursts of others rationally and constructively."

Writing this part is the most subjective of the three. It would probably be best to get an expert to assist you or to write it for you. Your Human Resources department often has the skills and abilities to do this for you.

Evaluation

Once a job description is written, the interview should be constructed around it. Systemaitec recommends that you develop a standard interview and evaluaton package for each craft. Anytime a craft position is to be hired or reviewed, this package should be used. This assures that every candidate/craftsperson will be asked the same questions and be evaluated against the same criteria. When the EEO comes knocking on the door to examine your hiring/evaluation practices, you need to be able to show them that you evaluate all candidates in an impartial and consistent manner. You can't ask one candidate one set of questions and another candidate some other set of questions.

Interview

One way to evaluate candidates/craftspersons is to interview them. This is especially important in evaluating the soft skills included in the Societal Abilities. Every candidate should be interviewed by a team of people. That team should consist of the department head, the foreman to whom the candidate would or does report, another foreman if one exists, and one like-craftsperson.

The interview should not take place as a panel discussion. At most, two people should interview the person at a time. The interview package should be broken up so that each group or person has a different emphasis. For example, the foreman and department head might evaluate Societal Abilities while the craftsman and the other foreman might work on Physical Abilities. They might split mental abilities between them. In any case, the strategy should include having similar questions for each interviewer to ask so answers can be compared.

Outside Skills Evaluation

There is no such thing as a good interview. Interviews take a snapshot of the candidate under the best of conditions. You cannot really tell if he/she will perform under real conditions. So, how do you get that information and sharpen your ability to pick the candidate who really is the best?

Many technical schools and community colleges are offering a relatively new service. If you provide them with the type of job description described above, they will test your candidates using federally approved test methods. They can test physical abilities, cognitive abilities, education, and teamwork skills. They provide a written report on each candidate with a numeric score.

For most firms, this is the best way to evaluate the candidates. It has been my experience that the margin of error in the tests is fairly large so that two candidates who score close to one another cannot be ranked by the scores. Still, it will weed out the people who simply cannot do the job or who have no desire to do it.

Initial Training

In many cases, you will find a qualified craft to fill your open position. In others, the occurrence of which is growing, you will only find the best of the unqualified to hire. Often you will have to train them yourself.

In either case, you must do some initial training. Only the very rare birds can come into your plant and be able to do everything you want them to do. You undoubtedly have some equipment that is simply unique to your process.

There are several ways to bring people "up to speed."

In-House classroom training

This is the most efficient but also the highest initial cost to train new people. You hire a trainer, put together lesson plans, and train the subjects that you feel are important to every new-hire. There are many craft trainers who will come to your plant and instruct for you on per student or per class basis. It seems inefficient because you have your new person collecting a full wage while sitting on their butts in a classroom. In reality, it is probably the quickest way to get information into them.

During this time, an overview of the process is taught so they understand what is going on; the basic safety procedures are reviewed; the basic workings of the management systems—how they will get work, how to read a work order, how to fill out a work order, what PM and PdM is done, etc., as well as the intricacies of your specialized equipment.

Local Technical School classroom training

This has some advantages, especially when you are training the one you hired for aptitude and attitude rather than existing skills. The employee gets professional training with a large class. You don't have to hire a trainer nor provide a training room and all of the specialized equipment (alignment kits, tools, mock-ups, etc.)

It has some disadvantages, also. Most likely, the school is not going to be able to train the new-hire on your specific equipment. You will end up having to do that training in-house anyway.

OJT

This is the favorite of most plants because they think they are getting more for their money. Actually, they are not only taking longer to train the new-hire, they are slowing down an existing employee while he/she tries to explain what they are doing to the trainee. Not only do you still have to pay the new-hire for the training time but you have to put up with lower direct work from the trainer.

OJT, though, has its place. It should be used in conjunction with classroom training. When you have hired for aptitude instead of developed skill, going to class one day, then working in the field the next can develop and reinforce needed skills. It also adds a sense of reality and urgency to the classroom setting.

Systemaitec recommends a combination of classroom and OJT training for all new hires.

Vendor Seminars or Training

Sometimes, the one who built your equipment will train your new-hire on that equipment for you. Most of the time this requires the payment of a fee. That fee can be very substantial as in the case of training on some instrument systems. Unfortunately, it is probably the only way to get your people trained sufficiently.

Vendor training is often available at your site. Check with the vendor of your equipment to see what they offer and what they charge. Some of the cheapest training I have offered to my crews has been to bring a technician in from the manufacturer and have my crew work with him during a rebuild of a complicated piece of equipment. As maintenance it is very expensive. As training it is very economical.

Development

Every craftsperson you have working for you needs refreshing on their skills and the learning of new skills as equipment evolves. If they only install a mechanical seal once a year, they are not going to be proficient at it. If they only weld every three months, they probably could use some refreshing before they pick up that stinger. If your maintenance is working well and you only open that complicated packaging machine every couple of years, then your crafts are going to be rusty when it comes to setting it back up.

All of the same options are available for development or refresher training as you have for new-hires. What is the best way? I can't tell you that until I see your system, your equipment and your craftspeople.

In general, the best way is to send them to vendor seminars or to the local tech school. Some companies have required their people to take a certain number of continuing education courses each year to keep up on the latest tools and techniques. A good maintenance manager will routinely bring in manufacturer representatives, salesmen, and other vendors to give short classes on their products.

One of the most productive training sessions I ever arranged as a maintenance manager was a seminar by the local SKF bearing representative. It lasted just two hours and was absolutely free. There was more practical information about the maintenance of bearings presented in that seminar than most people get in five years any other way. Bearing maintenance improved tenfold because of the tips and hints given by those sales reps.

Who pays?

To the craftspeople, this is the big question. There are advantages both ways. In some professions, such as nursing, continuing education is mandatory. Paying for it is sometimes the individual's responsibility and sometimes the employer pays. The best systems I have seen have been the ones that require the individual to pay but the company reimburses them upon successful completion of the course.

Should crafts be any different? Systemaitec recommends that you require some level of continuing education and offer tuition reimbursement for successful completion. In addition, you should be arranging seminars and in-house training sessions to meet some of that requirement.

MAINTENANCE SAFETY MODULE

It is possible to have a Lost Time Injury[1] rate of zero and a Recordable Injury[2] rate of less than one.

For many people who are struggling to keep their employees safe, this seems like pie-in-the-sky. Yet, there are plants, divisions and whole companies that are performing at that level. They have accomplished this through straightforward hard work. There are, though, some common elements to each program that all employers should implement.

For our purposes, we call this a maintenance program. It is Systemaitec's opinion that all maintenance departments should implement this module, even if the rest of the organization does not. In the best of all worlds, the whole organization would utilize these techniques so that everyone's safety improves.

---

1. A Lost Time Injury is defined as one that requires the employee to miss the start of his/her next normally assigned shift, or to miss subsequent scheduled work because of continued problems with the injury. The rate is the number of injuries per 200,000 man-hours of work.
2. A Recordable Injury is one that requires the ministrations of a doctor to treat or creates a 2nd degree burn over an area larger than a dime. Any period of unconsciousness and all Industrial Illnesses are considered recordable. The rate is the number of injuries per 200,000 man-hours of work.

Why would you want a good safety record? First, it is legally, ethically and morally the right thing to do. Operating an unsafe workplace is illegal and could cost you a lot of money in fines and very bad publicity. You as an employer have an obligation to create and maintain a safe work environment. Killing, maiming, or hurting people to increase your profits is simply wrong.

It is also short term and erroneous to think that working safely costs money or that working unsafely saves money. The cost of workers compensation insurance premiums and the loss or diminution of skilled labor will cost you much more than you will save by not implementing a good safety program.

Philosophy

Believe it or not, safety is not a program. Nor is it a process. At its heart, it is a philosophy or a value of the company. How management views and portrays safety is the foundation of everything else you do. If you do not believe that people should work safely but rather that they should take risks to save time and money, then your employees will take risks—and get hurt. They will do this no matter how many safety meetings you hold or how many prizes you give out.

I took a course on supervision many years ago in Philadelphia, PA. I was working for a local paper company at the time as a manager. In that course, the instructor said something that has stuck with me for thirty years: "Your employees will do what they think you want them to do." The key phrase there is "what they think." It doesn't matter what you tell them. Employees are constantly watching you to find out what you really want. You can talk safety for two hours a day. The first time you go into the plant and do something unsafe or instruct others to do so, you have blown your whole program.

So, the first part of this module is installing a safety philosophy, broadcasting it, and living by it. There are several you might want to consider. I don't suggest you adopt these verbatim but find the concept that seems to fit with your own values.

- "Safety is more important than anything else including profits."
- "All injuries are preventable and we are going to prevent all of them."
- "There is no amount of production that is worth hurting someone to get."
- "Working safely is a condition of employment. We will not tolerate unsafe acts."

These can get you started. Consider what your supervisors' philosophy really is. If it does not line up with the one you come up with, then it is up to you to change it. It must be a value for them if it is to be real for your employees.

Safety Policy

Once you have determined your philosophy or value about safety, you need to write a policy that explains that. The policy also states briefly how people participate in the safety process. Below is an example.

> *The maintenance department of the XYZ Company believes that each employee must think safety and be safe. Every employee is responsible for observing all safety rules all the time and insuring that all his/her co-workers do the same. Safe work is a job requirement and we will accept nothing less. All injuries, no matter how minor, are reported to a supervisor. All safety incidents, no matter how insignificant are reported on a safety incident form. Every injury or incident is investigated, a root cause found, and corrective action implemented to insure it cannot occur again.*

This policy should be printed up, given to all maintenance employees, and posted on bulletin boards where other employees will see it. It should be made conspicuous in the work place so that any infraction or failure to follow it is perfectly clear. Then, everyone in the department but must follow it consistently, especially the leadership.

Safety Meetings

Yes, this is a bug-a-boo. Most places hold some sort of safety meeting with the troops. Most of them are so boring and useless that they do more to harm safety than to help it. Some are given in such a cavalier fashion that it only reinforces for the employees that management does not really want them to work safe.

Systemaitec suggests three levels of employee safety meetings:

The Shift Starter

First, there should be a five to ten minute safety talk each morning. The maintenance supervisor should review all the jobs that are going on that day and cover the known hazards. If there is something special going on in the plant that could effect the department or any of the scheduled jobs, it should be mentioned at that meeting. This is done in the shop, standing up. Some refer to it as a toolbox meeting.

This meeting is best held just after making all the job assignments for the day and handing out the work orders. Before it is over, each craftsperson should be given the opportunity to ask any questions they might have about the job they have been assigned.

The Standing Agenda

This meeting should be held monthly. It should be for no more than one hour. This meeting is a formal meeting and attendance is mandatory. The meeting is broken up into several sections: an informational section, a feedback section, and training section. The length of each section can vary according to content.

In the informational section, the results of the safety effort are reviewed. All accidents are discussed with root causes explained and what corrective action is being taken. Ongoing safety projects are reviewed and updated. The status of all open safety suggestions is reviewed.

In the feedback section, the crew is asked to express safety concerns, make safety recommendations, and suggest changes to the program. All of the suggestions are recorded, action items are assigned to team members, and completion dates are agreed upon. The status of these would be reported in future meetings during the informational section.

In the training section, formal training on a particular safety topic is presented. This might be a pertinent video. There are many good videos out there for maintenance safety. Systemaitec suggests that you put together a library of them. Other topics might be: new MSA's (see below), new or modified procedures, health and hygiene, tool safety, driving safety, or training on the specific hazards of your site, such as noise, or dust, or electric arcs.

Systemaitec recommends a top-down approach to this meeting. The maintenance department head should hold a meeting with his direct reports. He/she should produce the lesson plan for the teaching section. The informational section should include all the information that his/her subordinates should be presenting to their meetings. The feedback section should record all the information from the individual crew meetings held the prior month.

The next level of management or supervision should then hold their monthly meeting. The same lesson plan should be presented in each meeting.

Minutes of all these meetings should be distributed to all attendees and filed by the department head for future reference.

A sample agenda form is in appendix B.

Annual Safety Extravaganza

Once a year, we suggest that you celebrate your accomplishments. I used to hold a donut and kolache breakfast for the entire maintenance department and selected guests. During the meeting we would review safety performance for the past year, set goals for the next year, and celebrate our victories. Normally, I would produce a humorous programmed text that everyone would work through before the end of the meeting. This is a great time to have plant management and production management join you in celebrating your crew's accomplishments.

Evaluation

As I noted before, a poorly done safety meeting can do more damage to the program than good. Systemaitec suggests that the maintenance department head attend one crew meeting a month and evaluate the performance of the meeting leader. This should be followed by an honest feedback session meant to improve the meetings. If the leader needs platform skills, technical skills, or people skills, training should be conducted as soon as practical. This should be made part of that leader's performance review.

Maintenance Inspections

One of the ways for management to show that they are serious about producing a safe work place is to require that Maintenance maintain its shop area and tools in a very clean, very safe manner. "You don't get what you expect. You get what you inspect."

Housekeeping

The Maintenance department should be clean. Sloppy work areas are not only dangerous, they are an indication of sloppy work habits. Simply insisting that the shop be kept neat and clean is the best way to send the message that you insist on good work habits in everything they do. If you are willing to "spend the money" to have a clean shop, you are sending the message that you are serious about safety.

Systemaitec recommends that you develop a set of cleanliness and neatness standards. Walk through every section of maintenance's area with a clipboard. Look at everything that you would inspect. Write down how you want it to look, e.g., the floor should be swept, free of trash, with no standing water, oil or grease, no tripping hazards. Worktables should be clean, free of trash, with no parts or fasteners left from previous jobs. Only current work should be on the table. There is no grease, oil or water on the tabletop. Window glass should be clean and free of cobwebs. No missing panes and no cracks. Etc.

When you have written the standards, they should be given to the crews with the explanation that this is your requirement for how the area should look all the time. Housekeeping inspections should be made part of your PM program, with someone conducting an inspection on a short-term periodic basis. On at least a monthly basis you should conduct an audit of the program. Walk through with the last inspection performed and insure that the area meets your minimum standards or that the deficiencies you find are recorded and scheduled for correction.

Tool and Equipment Inspections

Your supervisors should hold a periodic inspection of all hand tools. This means every craftsperson's box and all tools. Broken or modified tools should be eliminated and replaced with the correct tool.

All electrical tools should have a ground-fault check performed routinely. Those that pass should be marked as passing. Those that fail should be removed from the tool crib immediately. Systemaitec suggests that you use colored tape on the electric cord to mark the good equipment. Use a different color for each inspection cycle. Display the color code on a board in the tool crib so that anyone removing a tool knows when it was last checked good.

All rigging equipment should be inspected for serviceability periodically. Any sling, harness, or choker found defective should be removed immediately and discarded. All lifting equipment should be load tested periodically to insure it will carry the rated load.

All maintenance cranes and hoists, including overhead cranes, jibs, come-alongs, and chain hoists should be inspected routinely for serviceability. Any defect should result in a "red-tag" which indicates the equipment is not to be used. If it is an electrical, hardwired device, it should be locked out.

It just makes sense to do these inspections. What is more, it sends the right message to the crew: "I am not going to allow the use of substandard or dangerous equipment. I am willing to spend the money to keep that from happening. It is important enough that I am willing to use crew time to inspect it and fix it."

Personal Protective Equipment

Maintenance is dangerous. We work under, over, and around others who are removing large, heavy pieces of equipment, heavy fasteners, and pieces of steel.

We handle sharp objects, metal with burrs on it, greasy equipment, and process equipment that may be contaminated with dangerous chemicals. The chance of getting hurt in maintenance is greater than in any other profession besides construction.

So, why would we approach every day unprotected? If a fireman goes to a fire scene, he is dressed in bunker gear. When an astronaut gets on the space shuttle, he dresses in his space suit. When a maintenance person goes to the field to work on a piece of equipment, he/she should be dressed in his/her protective clothing also.

What PPE should your crew use? That depends on the conditions of your plant. There are, though, some basics that every crew should have.

Hard Hats

Does your crew work below equipment that is operating? Do they work below other crews? Do they walk through areas where there are structures above them? Then they should wear a hard hat all the time. That's crazy, you say? The chance of them getting hit on the head is very small. That's correct. If we could predict when an accident is going to take place, then we would only need PPE just as it happens. Unfortunately, an accident, by definition, is unpredictable. The only way to achieve the zero injury level is to protect them all the time. So, require that hardhats be worn all the time, except in the offices or in a vehicle.

Safety Glasses

Same philosophy as above. If we knew when something was going to fly into our eye, we could put the glasses on just as it happens. Until we get that good in our prognostication, insist that everyone wear safety glasses in all work areas.

Remember this: If you require safety glasses and there is the potential for flying particles in the work place, the glasses must be fitted with side shields.

Appropriate gloves

Maintenance almost always works with sharp, rough, or worn metal objects. Unlike cowhide, our hide is pretty soft and can be cut easily. Insist that your maintenance people put some cowhide between them and that work. It is the simplest way to reduce cuts that result in stitches—a definite recordable injury.

At other times, the appropriate glove may be cut resistant Kevlar, or chemical resistant rubber. You need to decide what gloves are needed on each job and insist that your people wear them.

Safety Goggles

If your employees work with any kind of liquids—corrosive or not—they should be required to put on goggles. If they work in a dusty environment, they should be required to wear goggles. That doesn't mean they should take the safety glasses off. The goggles will stop the dust and liquid but they may not stop a high-speed chip of steel. The glasses will.

Other PPE

You must look at every job and determine what PPE is appropriate. Slicker suits, steel-toed boots, rubber boots, leathers, leggings, respirators, SCBA's…all of these may be appropriate at times. You have to decide if there is a hazard on the job that can only be mitigated by the use of PPE then insist on the wearing of that gear when that job is done.

Getting people to use PPE when they have not had to in the past is a culture change. It should be handled like any other change. There must be a lot of communications about the change. The reasons for doing it must be clear. The benefits of the change must be communicated over and over. Listen to the feedback. Modify the plan where appropriate. Set a date for implementation and stick to it. After that date, insist that the PPE be worn every time. There should never be a reason accepted for not wearing it. If necessary, discipline the employee and the supervisor when an infraction occurs. Reward early compliance with recognition and praise. It will take months of consistent effort to make the culture change.

MHA

Maintenance Hazard Analysis is the process Systemaitec recommends for developing Maintenance Work Procedures. The effort that goes into this is intense. There is no department that could do all the MHA's needed immediately. Systemaitec recommends that you identify all the most hazardous tasks that your department does, prioritize them, then begin doing one MHA per week until they are all done. After that, MHA's could be done on all other tasks on a once per month basis or on an as needed basis.

A sample MHA form is in the appendix B.

The first step in MHA is to determine what safety equipment and PPE are required for the job. These are listed in the appropriate space on the form. This may change as the MHA proceeds but it is good to start with your first instincts.

The particular task is then broken down into distinct steps. These should be fairly detailed but not to the level of which screw to turn or wrench to use. For example, if the MHA was on replacing a defective pump, the steps might be:

1. Lock out the pump.
2. Inspect the pump to insure that the inlet and outlet valves are closed and locked out. Insure the drain valve is open and that there is no liquid left in the pump housing.
3. Insure the barrier fluid re-circulator is shut off.
4. Insure the barrier fluid has been drained.
5. Remove the coupling guard.
6. Break the coupling.
7. Remove the seal fluid tubes from their connections on the housing.
8. Remove the bolts holding the housing to the volute.

Etc.

Each step is then walked through mentally. Include craftspeople in this step to insure accuracy. Any potential hazard associated with that step is listed in the hazard section of the form. An example of this might be:

Step	Hazard
5. Remove the coupling guard	Sheet metal guard may have rough or sharp edges. Guard may have chemical contamination. Access to the guard is difficult. There is a back strain hazard if done improperly. This is a dusty environment. Airborne particles may be present.

When all steps have been processed in this way, return to the first step and analyze each hazard. In the third column, list ways to avoid or mitigate the hazard. This might include PPE, using a particular tool, removing some other piece of equipment before starting that step, or shutting off or locking out more equipment, valves or pipes.

In our last example, it might look like this:

Step	Hazard	Mitigation
5. Remove coupling guard	Sheet metal guard has sharp edges	Wear leather gloves at all times.
	Guard may have chemical contamination	Wash area thoroughly prior to touching guard. Inspect inside of guard for possible contamination. Where chemical resistant gloves.

This would continue until every hazard is addressed.

This information is filed for each job. Using this information, write a Maintenance Work Procedure for this task. This can be written in two-column format: column one is the name of the step. Column two contains the actions to be taken including the mitigation steps you found in the MHA. It can also be written in paragraph format. What format you use is personal preference.

This is more detailed than the MHA. The steps are broken down more finely. Tools and parts are listed. When complete, you have a procedure for doing this job that informs the reader of how to do the whole job safely. Along with the MHA, it can provide valuable information to a new member of the crew or can be used to train and refresh current employees.

Incident Reporting and Investigation

Reporting, investigating, and correcting failures is as important to improving your safety process as it is in improving your production processes. No system works perfectly all the time. The one that continuously improves, though, is the one that has a feedback and correction system built in. That is the purpose of Incident Reporting and Investigation.

There are two reports and one form that should be in your pack of feedback and corrective action tools. (See the report form in Appendix B)

The first is the injury report. Anytime someone is injured in your department, you should use this report form. An injury is anything that requires first aid or more extensive treatment. A Band-Aid is first aid.

The second is a safety incident report. A safety incident is anytime something occurs out of the ordinary that could have resulted in an injury but, by chance, didn't.

The supervisor of the injured employee should fill out the form. Record all pertinent information, visit the site of the incident, take pictures or make sketches if the incident was serious. Do everything you can to gather enough information to determine the cause or causes of the incident.

When a root cause is uncovered, record it in the appropriate section.

Ask yourself, what could be done to prevent this from happening again? The answers become action items and are recorded on the bottom portion of the form. The action items need to be assigned and commitments made to complete them. This information should be recorded in a manual or computerized database.

Safety incident investigation is a detailed process that is not part of the basic modules. There are many consultants that can train your people in these techniques. Systemaitec can recommend some good training in this area.

Accident and Incident Prevention and Observation Program

The acronym for this system is AIPOP. It is pronounced "Eye-Pop." The program is named that because situations that could cause a safety incident should cause your employee's eyes to pop out of their head. This is a hazard and problem recognition program. There are many of them on the market. Most of them will work if they are used appropriately and for the right reasons.

The AIPOP program is basically a reporting and correction system. Your employees are given easy to carry forms that allow then to write up a quick report when they see an unsafe action or condition. The real value of this program is that it requires those who find the problem to also correct it.

That may seem like a big task—and it would be if this was the only safety incident reporting system you were using. This system, though, is used in conjunction with Safety Incident Reporting and Injury Reporting. AIPOP concentrates on those things that occur or exist prior to an incident occurring.

Here is chart of how and when each process is used:

Type of occurrence	Program
Doctor Care	Injury Reporting
First Aid	Injury Reporting

Type of occurrence	Program
Near Miss—no injury	Incident Reporting
Property Damage—no injury	Incident Reporting
Unsafe Act—no incident	AIPOP
Unsafe Condition—no incident	AIPOP

As you can see, the systems complement each other. Missing any one of them leaves your program with a large hole.

What ever programs and processes you adopt, be sure that you are reaching far enough back into the process of an accident occurring to catch them before they start. Injury reporting is a lagging indicator. It can be advantageous to a new safety program and can fix problems after they have caused an injury—but a good safety program has to be aimed at preventing the injury to begin with. Incident reporting is a good second step. It is aimed at fixing problems that caused an incident to occur so that that situation can never end up causing an injury in the future. The problem is, you still let an incident occur before applying corrective action. By definition, that incident did not cause an injury by dumb-luck. To stop injuries, you have to prevent incidents. To stop incidents, you have to get to the problem before an incident occurs. This is what AIPOP or the other programs like it attempt to do.

Appendix A

▼

YOUR COMPANY NAME
Maintenance Work Request

| MWR Number: | | Prod. Code | Maint. Code | Priority |

Eq. Code | Eq. Name

Account# | Date of Request:

Symptom:

Failure Code:
1. Structural failure ☐
2. Bearing noise ☐
3. Gear noise ☐
4. Heat ☐
5. Vibration ☐
6. Leakage ☐
7. Seal failure ☐
8. No power ☐
9. Catastrophic ☐
10. Erratic ☐
11. Instrument ☐

Requested Action

Requested by: | Authorized by:

Shaded portions by requesting organization

Planning Section

Tools and Equipment | Parts and Supplies

Corrective Action Section

Corrective Action Taken:

Date work started: | Craftsman: | Date work completed:

- 79 -

The Seven Modules of Systematic Maintenance

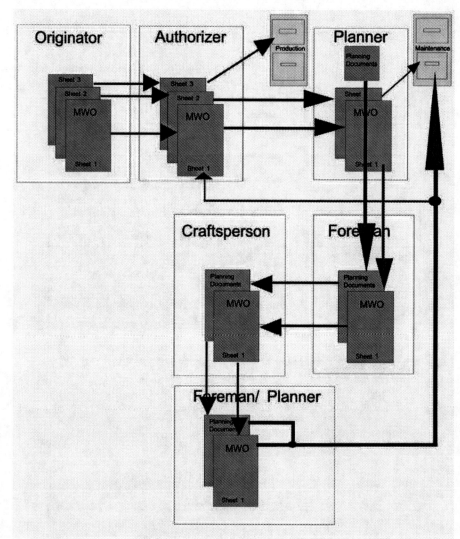

MWO Work Flow

Maintenance Work Request Priority System

Production Code

Code Description
10 Repairs are necessary to correct a potential or existing safety or environmental hazard
9 Breakdown of critical production equipment has occurred. The department is down.
8 Breakdown of critical production equipment is imminent. The department is running.
7 Equipment needs to be repaired to correct a real or potential quality problem. Preventive and Predictive maintenance work.
6 Repairs are necessary to ensure compliance with production schedule.
5 Work to increase onstream, or modify the equipment to increase production or quality.
4 Repairs to on-line spares or peripheral equipment.
3 Discretionary cleaning, painting, or fix-up.

Maintenance Code

Code Description
10 Equipment breakdown has occurred or is imminent
9 Repairs will prevent or correct a potential future safety or environmental hazard
8 PM, PrM
7 Scheduled overhauls of non-spared equipment, replacement of production or critical utility equip.
6 Rebuild of critical spares units, including inline spares.
5 Rebuild of non-critical spare units including spares in the stores system
4 Repair of maintenance and production tools.
3 Non-capital modifications of equipment and facilities to meet minimum standards. Major maintenance work not already covered.
2 Implementation of the painting and protective coating major maintenance plan.
1 Discretionary cleaning, painting or fix-up

Appendix B

Your Company's Name
Maintenance Hazard Analysis

Equipment Number:	Equipment Name:	Date of Review

Analysis Team Members:

Name	Name	Name	Name

Title:

Description of Task:

PPE Required	Safety Equipment Required:

Step	Hazard	Mitigation
1.		

	YOUR COMPANY'S NAME	
Date of Report	**Safety Incident Report**	Report taken by:
Incident Number:	Incident Title:	
Location of Incident:		Equipment # Involved:

Description of Incident:

Description of Injuries or damage:

Investigation Team Members:

Name	Name	Name	Name

Root Cause:

Contributing Causes:

Corrective Action Items:

Item Number	Action	Commitment	PPR

Lead Investigator		Investigation Approved by:	
	Date:		Date:

YOUR COMPANY'S NAME						
Date:	colspan="4"	Monthly Safety Meeting Agenda		Crew:	
Lead by:							
Attendee	Init	Attendee	Init	Attendee	Init

Informational Section

Review of Safety Incidents and Action Items: (attach copies of report)

Incident #	Incident	A/I	Status

Review of Safety Suggestions: (attach copies of report)

Suggestion #	Suggestion	Status

Review of Safety Projects

Project or WO	Title	Status

FeedBack Section

Feedback, comments, Safety Suggestions   PPR   Commitment

1.

2.

3.

4.

Training Section

Topic:
Method(s) of Delivery: Lecture   Video   Computer   Text   other: _____ (circle all that apply)
Summary of information: _____

0-595-30421-4

CPSIA information can be obtained at www.ICGtesting.com
Printed in the USA
LVOW05s2055250713

344656LV00001B/187/A